Springer Proceedings in Mathematics & Statistics

Volume 158

Springer Proceedings in Mathematics & Statistics

This book series features volumes composed of selected contributions from workshops and conferences in all areas of current research in mathematics and statistics, including operation research and optimization. In addition to an overall evaluation of the interest, scientific quality, and timeliness of each proposal at the hands of the publisher, individual contributions are all refereed to the high quality standards of leading journals in the field. Thus, this series provides the research community with well-edited, authoritative reports on developments in the most exciting areas of mathematical and statistical research today.

More information about this series at http://www.springer.com/series/10533

M'hamed Eddahbi · El Hassan Essaky
Josep Vives
Editors

Statistical Methods and Applications in Insurance and Finance

CIMPA School, Marrakech and Kelaat M'gouna, Morocco, April 2013

Springer

Editors
M'hamed Eddahbi
Department of Mathematics
Université Cadi Ayyad
Marrakesh
Morocco

Josep Vives
Faculty of Mathematics
University of Barcelona
Barcelona
Spain

El Hassan Essaky
Faculté Polydisciplinaire de Safi
Université Cadi Ayyad
Safi
Morocco

ISSN 2194-1009 ISSN 2194-1017 (electronic)
Springer Proceedings in Mathematics & Statistics
ISBN 978-3-319-80804-8 ISBN 978-3-319-30417-5 (eBook)
DOI 10.1007/978-3-319-30417-5

Mathematics Subject Classification (2010): 91B30, 60H07, 60H35, 60G51, 60E07, 60J65, 60G52, 62P05, 91B28, 60H30, 91G80, 91G20, 90B30, 60J75, 62M10, 60H10, 60G44, 93E20, 60G55, 60H20, 60H99

Printed on acid-free paper

This Springer imprint is published by Springer Nature
The registered company is Springer International Publishing AG Switzerland

The original version of the book frontmatter was revised: Book subtitle was corrected. The erratum to the book frontmatter is available at DOI 10.1007/978-3-319-30417-5_10

Preface

This book is an outcome of the CIMPA-UNESCO-MESR-MINECO-MOROCCO research school entitled "Statistical Methods and Applications in Finance and Actuarial Science". The research school, organized by the Cadi Ayyad University in Marrakech, in collaboration with the International Centre for Pure and Applied Mathematics (CIMPA), was held in Marrakech and Kelaat M'gouna between 8 and 20 April 2013.

This volume of proceedings from the conference provides an opportunity for readers to engage with the lecture notes for two of the courses and seven refereed papers that were presented during the school.

The volume comprises two parts. The first is devoted to applications in Finance and includes a series of lectures presented by F. Viens during the conference entitled "A didactic introduction to risk management via hedging in discrete and continuous time" as well as three refereed contributions. The first of these, by M. Eddahbi and S.M. Lalaoui Ben Cherif, entitled "Sensitivity analysis for time-inhomogeneous Lévy process: a Malliavin calculus approach and numerics", is devoted to the study of sensitivity analysis, with respect to the parameters of the model, within the framework of a time-inhomogeneous Lévy process. The second, by N. Privault and D. Yang, is entitled "Variance GGC asset price models and their sensitivity analysis" and treats the problem of computation of sensitivities or Greeks under different examples of Lévy type models. On the other hand, the third contribution, by J. Vives, entitled "Decomposition of the pricing formula for stochastic volatility models based on Malliavin-Skorohod type calculus", treats the problem of obtaining decompositions of the derivative price formula for stochastic volatility jump diffusion models that clarify the exact role of correlation and jumps in derivative prices.

The second part of this volume is devoted to applications to Insurance and the study of stochastic differential equations of different types. This part opens with the lecture notes for the course by B. Djehiche, entitled "Statistical estimation techniques in life and disability insurance—a short review". This lecture is a short introduction to some basic aspects of statistical estimation techniques known as graduation techniques in life and disability insurance. Subsequently, four refereed contributions are included. The contributions by A. Al-Hussein, A. Al-Hussein and B. Gherbal, entitled respectively "Necessary and sufficient conditions of optimal control for infinite dimensional SDE" and "Sufficient conditions of optimality for forward-backward doubly SDEs with jumps", are devoted to optimal control problems. The contribution by M. Benabdallah, S. Bouhadou and Y. Ouknine, entitled "On the pathwise uniqueness of solutions of one-dimensional stochastic differential equations with jumps", treats the problem of uniqueness of the solution of one-dimensional stochastic differential equations with jumps, and finally the contribution by E.H. Essaky and M. Hassani, entitled "BSDE approach for Dynkin game and American game option", is devoted to study of the existence of a value as well as a saddle point for a Dynkin game under weaker conditions. This contribution also discusses American game option pricing problems in finance and their relationship with backward stochastic differential equations with double reflecting barriers.

Marrakech
Safi
Barcelona
November 2015

M'hamed Eddahbi
El Hassan Essaky
Josep Vives

Contents

Part I
Finance

A Didactic Introduction to Risk Management via Hedging in Discrete and Continuous Time

Frederi Viens

Abstract The following is based on a series of lectures which the author presented at a graduate training workshop in April 2013, organized by the Ecole nationale des sciences appliquées of the Université Cadi Ayyad in Marrakech, Morocco, and partially financed by the CIMPA (International Center for Pure and Applied Mathematics). The author expresses his gratitude to the CIMPA and the main organizers Profs. M'hamed Eddahbi, Khalifa Es-sebaiy, Youssef Ouknine, and Josep Vives, for their work, support, and hospitality. The style of these notes is deliberately informal and didactic, with no formal development of a full mathematical theory. The intended audience includes finishing undergraduate students (3 years of college) and first year graduate students (4 years of college), with some basic background in calculus, linear algebra, differential equations, and probability. No prior knowledge of investment finance or actuarial science is required. No references are provided in the text. An excellent further treatment of many of the topics listed herein can be found in a book currently recommended by the Society of Actuaries for its treatment of "Financial Economics": Robert L. McDonald: *Derivatives Markets* (3rd Edition, 2012), Pearson Series in Finance.

Keywords Risk management · Option pricing · Overnight profit · Market making · Actuarial mathematics

Mathematical Subject Classification 2010 91G20 · 91B30 · 60J65

F. Viens (✉)
Purdue University, West Lafayette, IN, USA
e-mail: viens@purdue.edu

M. Eddahbi et al. (eds.), *Statistical Methods and Applications in Insurance and Finance*, Springer Proceedings in Mathematics & Statistics 158,
DOI 10.1007/978-3-319-30417-5_1

1 Introduction: Basic Insurance Question (Casualty)

Assume a home value is \$100,000, and the chance of it burning down is $= 0.01$ (1 %). Also assume that an insurer insures homes which are independent and identically distributed.[1]

Therefore the expected loss per house $= 10^5 \times 10^{-2} = \$1{,}000$. Indeed, the loss variable is $100{,}000X$ where X is a Bernoulli r.v. with parameter $p = 0.01$ (i.e. X equals 1 with probability 0.01 and 0 otherwise)

→ Pricing of the insurance claim needs to cover this expected loss plus the cost of running an insurance business (e.g. employee salaries), minus the interest earned on clients' premiums.

→ But what if major a event occurs? The insurance company would need more cash to cover large losses.

1. Company must **hold capital reserves**: The Central Limit Theorem says the more i.i.d. clients one holds, the less reserve per capital one might need. Indeed:
   Let $X^{(N)} := 10^5(\sum_{i=1}^{N} X_i)\frac{1}{N}$: here X_i i.i.d. Bernoulli (parameter $p = 0.01$), thus $X^{(N)}$ is the per-capital reserve. We have of course $\mathbf{E}\left[X^{(N)}\right] = 10^5 pN/N = 1{,}000$. However,

$$\sqrt{Var(X^{(N)})} = 10^5\sqrt{Np(1-p)}\frac{1}{N} = \frac{1000}{\sqrt{N}}$$

   which shows that the spread of the per-capita reserve decreases like constant/$\sqrt{N}$. More specifically, to be 95 % sure that one can cover all losses, we can look for the amount of per-capita reserve ε beyond the average per-capita reserve of \$1,000 such that the chance of the actual per-capita loss exceeding that level is 5 %: find ε such that

$$\begin{aligned}
& P(X^{(N)} > 10^3 + \varepsilon) = 0.05 \\
\iff & P(\frac{X^{(N)} - 1000}{10^4\frac{1}{\sqrt{N}}} > \frac{\varepsilon}{10^4\frac{1}{\sqrt{N}}}) = 0.05 \\
\iff & P(N(0,1) > \frac{\varepsilon}{10^4\frac{1}{\sqrt{N}}}) = 0.05 \text{ approximately, by the CLT} \\
\iff & \frac{\varepsilon}{10^4\frac{1}{\sqrt{N}}} \simeq 1.645 \text{ approximately, using a normal table} \\
\iff & \varepsilon = \frac{16450}{\sqrt{N}}.
\end{aligned}$$

[1]This could be an abusive assumption if an insurer insures all the homes in a given high-risk area, such as a coastal flood plain or a town in the U.S. midwest in an area which is prone to tornados.

We conclude that

- If we have 10,000 customers, then excess reserve needed per customer equals $\varepsilon = \$164{,}5$.
- But if we have 1,000,000 customers, then this excess reserve decreases to $\varepsilon = \$16{,}45$.

  Hence the use of aggregating as many customers as possible, to take advantage of this phenomenon of risk diversification (as long as our customers are i.i.d).

2. However, there is another way to manage risk: **Use Reinsurance.**
   Ask a reinsurance company to take on the risk associated with very large events only. If the total value $Claim$ of all claims exceed a certain level K, the reinsurer pays the insurer $Claim - K$ to cover those claims in excess of the large value K; otherwise the reinsurer pays nothing: thus the reinsurer pays

   - $\max(Claims - K, 0)$: This is the payoff of a call option where the asset = total claims and the strike price = K = level where reinsurance kicks in.
   - This contract can also be thought of as a put option for the insurer: the insurer has the right to sell to the reinsurer all the contracts that lost money: thus insurer may sell to reinsurer the negative quantity $-Claims$ if that amount is less than $-K$; hence the contract payoff is $max(-K - (-Claims), 0)$.

   In any case, there is a need to **price** this **payoff**, i.e. this **contingent claim** $\max(Claims - K, 0)$.
   Question: Can the reinsurance use the same method of pricing excess reserves ε per client, as the insurance company does with its own individual clients? Here the reinsurer's clients are individual insurance companies. Therefore...
   Answer is **Typically NO**: indeed the typical number of clients M for the reinsurer is never as large as $N = 10{,}00{,}000$, so can't rely on "diversification of risk", the reinsurer cannot use the CLT because the number of insurance companies (or contracts) M for a reinsurer is usually too small.
3. Need a **new pricing method** for reinsurance: **Hedging**.
   This method would be common to both reinsurance and financial derivatives markets; it is the method of derivatives (e.g. options) pricing.
   The word "Claims" can be replaced by the more generic term "value of a risky asset or index" ; this asset could be a stock price $S = \{S(t); t \in [0, T]\}$.
   Basic idea of hedging: a *market maker* sells a call option with strike price K, payoff $C_T = \max(S(T) - K, 0)$.

   - **Question**: can we try to hedge this payoff ahead of time by investing in the stock S and in a risk-free account with short rate r?
   - **IF** the answer is **"Yes this can be done perfectly"**, then the value of the call option at any time t prior to maturity T ($t \leq T$) is exactly the value of the hedging investment (portfolio). In the language of insurance, this value is the **premium** of the call option at time t, the value one would pay to buy the claim at time t.

Equivalently, at time $t = 0$ (say), one only needs to make an investment equal in value to the hedging portfolio, and rebalance the portfolio over time so that its value remains equal to what it is supposed to be at any time $t \leq T$, then this portfolio will be exactly equation to the payoff $C_T = \max(S(T) - K, 0)$ at time T.

- **Answer** to the question:
  Answer is Yes, in discrete time: there's a perfect answer (perfect hedging portfolio) using the binomial model.
  However, typically, the binomial model work well for time step $h = 1$ day, but is much to show for very liquid asset high frequency ($h = 5$ min). For the HF question, there is a perfect continuous time theory.
  Answer is Yes, in continuous time: the Black-Scholes model also leads to a perfect hedging portfolio, but one must be allowed to trade continuously.

4. **In practice**: one typically uses a continuous-time model such as Black-Scholes, but one only follows its hedging portfolio discretely in time; this discretization leads to **hedging errors**.

   - In other words, *hedging in practice is never perfect.*
   - Hedging errors can be large if a large asset price swing occurs in a short period.
   - The market maker may try to *immunize* her position against such risks. One way to do this is to buy a financial derivative that is related to the one she sold. We will see below that if we sell a call option, we can buy a certain amount of another call option to cover some of the risk, using a procedure called *delta-gamma hedging*.

In the sequel we will use the following numerical values to illustrate the three types of basic hedges (binomial, continuous-time Black-Scholes, discretized version of continuous-time Black-Scholes), as well as delta-gamma hedging (also a discrete hedge for continuous-time Black-Scholes):

- Stock S, $S_0 = 40$.
  - Stock's volatility is $\sigma = 0.3$.
- Time step $\Delta t = h = \frac{1}{365}$ = one day.
- We sell the call option $C_{40}(0, S_0)$ with
  - Strike price $K = 40$ ("at the money").
  - Maturity $T = \frac{91}{365}$ (3 months).
- Short rate $r = 0.08$.

2 Binomial Option Pricing

The most general one-period model for option pricing is a risk-free asset whose value at time 0 is $B_0 = 1$ and at time h is $B_h = e^{rh}$, plus the following stock model:

new stock value if it goes "up": S_0u option payoff $= C_u$

↗

intial stock value: S_0

↘

new stock value if it goes "down": S_0d option payoff $= C_d$

For example, for a call with strike price K, we have $C_u = \max(0; S_0u - K)$ and $C_d = \max(0; S_0d - K)$. Here u and d are fixed values, and we assume $d < u$; using $d > 1$ or $d \leq 1$ are both legitimate, as long as we have $d < e^{rh} < u$ which is needed to avoid arbitrage. Note that we did not specify the probability of the stock going up or down; these so-called *objective* probabilities are not needed to price and hedge the option.

- **Hedging question**
  Find a portfolio $\ell = (b, y)$ with y shares of S at time 0 and b dollars in risk-free asset at time 0, such that value $V^\ell(h)$ at time h = exactly C_u if stock went "up" and C_d if stock went " down". Therefore we have the following values for the portfolio at times 0 and h

$$V^l(0) = b + yS_0,$$
$$V^l(h) = be^{rh} + yS_h = \begin{cases} be^{rh} + yS_0u \text{ if stock went "up"} \\ be^{rh} + yS_0d \text{ if stock went "down"} \end{cases}.$$

  To have a perfect hedge, only need to require that we *replicate* the option, i.e.

$$\begin{cases} be^{rh} + yS_0u = C_u \\ be^{rh} + yS_0d = C_d \end{cases}.$$

  This is a system with two unknowns b and y. and a unique solution (a **perfect hedge**)

$$y = \frac{C_u - C_d}{u - d}; \quad b = e^{-rh}\frac{uC_u - dC_d}{u - d}.$$

- **Pricing question**
  The price of the option at time 0 should be the value $V^\ell(0)$ of the hedging portfolio ℓ at time 0:

$$\text{Price of option at time } 0 = V^\ell(0) = b + yS_0$$

  with the values y and b given above.

- **Probabilistic interpretation of the option price**
  Let $q_u = \dfrac{e^{rh} - d}{u - d} \in (0, 1)$; then we find, after some simple algebra, that

$$V^l(0) = e^{-rh}(q_u C_u + (1 - q_u)C_d).$$

We can interpret this as saying that $V^\ell(0)$ is the discounted expected value of the option payoff at maturity (at time h) under a model in which the probability of going up is q_u, and therefore the probability of going down is $1 - q_u$: the payoff is

$$\text{payoff at time } 1 = \mathscr{X}_h = \begin{cases} C_u \text{ with prob } q_u \\ C_d \text{ with prob } (1 - q_u) \end{cases}$$
$$\text{price at time } 0 = e^{-rh}\mathbf{E}^{\mathbf{Q}}[\mathscr{X}_h].$$

Here $\mathbf{Q}$ is the probability measure defined by $\mathbf{Q}$("up") $= q_u$ and $\mathbf{Q}$("down") $= 1 - q_u$.
- $\mathbf{Q}$ is called the *risk-neutral measure* for our model. Notice that $e^{-rh}\mathbf{E}^{\mathbf{Q}}[S_h] = S_0$, which explains why $\mathbf{Q}$ is also called a *martingale measure*. The term "risk-neutral" comes from the fact that the strategy to hedge the option does not take into account the true risk associated with the stock S (e.g. its true chance of going up), and therefore this strategy is neutral with respect to the stock's risk. The formula $e^{-rh}\mathbf{E}^{\mathbf{Q}}[\mathscr{X}_h]$ is called the *discounted risk-neutral valuation formula* for the option price.

IMPORTANT: $V^l(0)$ is the correct ("arbitrage-free") price of the option C at time 0.

Also (easy to prove): the condition $q_u \in [0,1] \Leftrightarrow d < e^{rh} < u \iff$ no arbitrage $\Leftrightarrow$ there is a risk neutral measure.

Also: $\exists$ hedge $(b, y) \iff \exists!$ risk-neutral measure.

3 Multi-period Binomial Model: N Periods

To extend the binomial model to several periods, in an effort to develop a model for option pricing and hedging which includes the possibility of dynamic portfolio allocation, we consider a total number of periods $N \geq 2$, and iterate the one-period construction of the previous section, over several periods, forming what is known as a binomal tree, with the root typically represented at the left, and the leaves at the right, i.e. with time running from left to right. For $N = 2$, this tree representation for the two-period binomial model has the following form

$$
\begin{array}{ccccc}
 & & & & S_0u^2 \\
 & & & \nearrow & \\
 & & S_0u & & \\
 & \nearrow & & \searrow & \\
S_0 & & & & S_0du \\
 & \searrow & & \nearrow & \\
 & & S_0d & & \\
 & & & \searrow & \\
 & & & & S_0d^2
\end{array}
$$

Note that the tree has the so-called "recombining" property, because the up and down factors do not change. More generally, for $N \geq 2$, and $n = 1,2,\ldots,N$, $k = 0,1,2,\ldots,n$, period number n models the dynamics during the time interval $[n-1,n]$, and the node (n,k) is the name of the position in the tree at time n for any stock price path which takes k up steps and $n-k$ down steps, from time 0 to time n. This parametrization of nodes is only possible because of the recombining property. This property works because the up and down factors u,d do not depend on the position in the tree at a fixed time (u and d might depend on time n, this does not impact the recombining property). In particular, the value of S at node (n,k) is

$$S_{n,k} := S_0 u^k d^{n-k}.$$

Again in the case where u and d depend only on n (we will not consider other cases in these notes), let us translate the binomial model in a more probabilist fashion. Let q_u be probability to go up at every node. Assume q_u is constant. For every $n = 1,\ldots,N$, we can consider the random variable K_n representing the number of times that the stock went up rather than down between time 0 and time n. Then $K_n = \sum_{i=1}^{n} \varepsilon_i$ where $\varepsilon_i = 1$ if the stock went up in the interval $[i-1,1]$, and $\varepsilon_i = 0$ if the stock went down. Each ε_i is a Bernoulli random variable with parameter q_u. Assuming all the ε_i's are independent, K_n is thus a binomial random variable with parameters n,q_u. This is from whence the binomial model gets its name. The stock price model then has the following probabilistic representation: for $n = 0,\ldots,N$,

$$S_n = S_0 u^{K_n} d^{n-K_n} = S_0 e^{(\ln d)n + \ln(u/d)K_n}.$$

4 Option Pricing and Hedging Algorithm: Backwards Recursion

Assume we need to price and hedge a simple European claim with contract function Φ. We may use the pricing and hedging scheme from the one-period model iteratively backwards in time to price and hedge options in the N-period binomial model.

- At time N, value of option = value of payoff = $C_N = \Phi(S_N)$
  In node notation, if number of up steps $= k$ then $S_N = S_0 u^k d^{N-k}$. Thus:

  Initialization $\quad C_{N,k} = \Phi(S_0 u^k d^{N-k})$ for all $k = 0, 1, \ldots, N$.

- To implement the recursion, assume that at some time $n \le N$; $C_{n,k}$ is known for every $k = 0, 1, \ldots, n$. Then, for each $k = 0, 1, \ldots, n-1$, by using the formula for pricing in the one-period binomial model which goes from node $(n-1, k)$ to the two possible future nodes (n, k) and $(n, k+1)$, i.e. the tree, in which each node contains stock and option prices,

$$
\begin{array}{ccl}
 & & \begin{matrix} S_{n,k+1} \\ C_{n,k+1} \end{matrix} \quad \text{with probability } q_u \\
 & \nearrow & \\
\begin{matrix} S_{n-1,k} \\ C_{n-1,k} \end{matrix} & & \\
 & \searrow & \\
 & & \begin{matrix} S_{n,k} \\ C_{n,k} \end{matrix} \quad \text{with probability } (1-q_u)
\end{array}
$$

  we obtain the following price recursion:

  Recursion: option prices $\quad C_{n-1,k} = e^{-rh}\left(q_u C_{n,k} + (1-q_u) C_{n,k-1}\right)$ for each $n = 1, \ldots, N$ and $k = 0, 1, \ldots, n-1$.

- We must also compute the hedging portfolio at node $(n-1, k)$: this can either be computed while implementing the previous recursion, or offline after all option prices are known. By the one-period hedging portfolio, this is

  Hedge: number of shares of stock

$$y_{n-1,k} = \frac{C_{n,k+1} - C_{n,k}}{S_{n,k+1} - S_{n,k}} = \frac{C_{n,k+1} - C_{n,k}}{S_{n-1,k}(u-d)}.$$

  Hedge: wealth in risk-free account

$$b_{n-1,k} = C_{n-1,k} - S_{n-1,k} y_{n-1,k}.$$

- The above scheme provides a perfect hedge which can be followed dynamically in time, and reacts to the changes in stock prices over each period. Indeed, at time $n-1$, the hedging decision only requires knowledge of the observed stock price $S_{n-1} = S_{n-1,K_{n-1}}$ (under the binomial model, the random variables S_{n-1} and K_{n-1} can be computed one from the other; they share the same information), and of the two possible future values for C_n, which are $C_{n,K_{n-1}+1}$ and $C_{n,K_{n-1}}$ which are among the precomputed values in the binomial tree.

- As an option hedger (also known as an option market maker), one must consider the trade-off between assuming that the length of the time in one's binomial model is short enough to stick closely to the stock variations, and incurring many transaction costs every time one rebalances one's hedging portfolio. In practice, stock prices change many more times than once a day. Yet market makers often assume that $h = 1/365 = 1$ day nonetheless. When we enter our discussion of continuous-time modeling with discrete-time hedging, we will provide a way to compute the discrepancy between a perfect hedge and the need to keep the hedging frequency down to a reasonable, daily level.
- Recall probabilistic representation of the option price for one period: price at time $0 = e^{-rh}\mathbf{E}^{\mathbf{Q}}[\mathscr{X}_h]$. Such a formula also holds for the multiperiod model, and it is easy to prove this by using the one-period formula and the recursion formula for the multi-period model given here. The details are left to the reader.

Discounted Risk Neutral Valuation Formula: multi-period case With maturity $T = Nh$, define the contingent claim $\mathscr{X}_T = \Phi(S(T))$, the risk-neutral (martingale) measure $\mathbf{Q}$ is defined by using the risk-neutral probabilities q_u and $1 - q_u$ in each period. Then we have

$$\text{price of } \mathscr{X}_T \text{ at time } 0 = e^{-rT}\mathbf{E}^{\mathbf{Q}}[\mathscr{X}_T].$$

4.1 How to Estimate/Calibrate Parameters r, u and d?

We provide some brief recommendations for the parameter estimation question.

An excellent proxy for the rate r is the LIBOR (London Interbank Offered Rate) short (overnight) rate L: this is the average rate at which banks lend each other money over a 24-hour period. This is thus most appropriate when $h = 1/365$, and one sets $e^{rh} = 1 + L$. Since the LIBOR short rate changes over time, one typically uses the previous day's value of L to calibrate r. There exist stochastic models of interest rates that take into account the uncertainty on future values of L. They are not discussed in these notes.

For u and d, it is typical to base their estimation/calibration using the concept of "historical volatility", which can be defined, for instance, as the empirical standard deviation, over an appropriately long time period, of $h^{-1/2}\mathscr{R}_t$ where $\mathscr{R}_t$ are the log returns $\mathscr{R}_t := \log(S(t-h)/S(t))$ where $(S(t))_t$ are the past stock price data, which can thus be identified, insofar as it represents a consistent estimator, with the square root σ of

$$\sigma^2 := Var(\mathscr{R}_t)$$

where now the notation $\mathscr{R}_t$ comes from a specific model, as long as this variance does not depend on t (stationarity).

There are many other ways of determining volatility models, some of which involve assuming that volatility is random itself. An emerging method for determining volatility is becoming popular in the case of the S&P 500 index. Since 1993, the Chicago Board of Options Exchange (CBOE) has published a composite value of option prices on this equity index, which can be interpreted as a 30-day average of volatility on the index. This volatility index, now known as the VIX, increased in popularity since the CBOE started offering traded options and futures contracts on the VIX starting in 2004.

Once a value of σ has been determined, a common method for calibrating u and d using the so-called Cox-Ross-Rubenstein parametrization:

$$u = e^{\sigma\sqrt{h}},$$
$$d = e^{-\sigma\sqrt{h}}.$$

Other choices include the Hull-White parametrization $u = 1 + \alpha h + \sigma\sqrt{h}, d = 1 + \alpha h - \sigma\sqrt{h}$; as well as the Jarrow-Rudd parametrization $u = e^{\mu h + \sigma\sqrt{h}}, d = e^{\mu h - \sigma\sqrt{h}}$, where the values of α and μ are h^{-1} times the expected values of the log returns $\mathcal{R}_t$ or the simple returns $R_t = (S(t+h) - S(t))/S(t)$.

Understanding the differences between these various parametrizations can be done in conjunction with the introduction of the continuous-time analogue to the Binomial model, the so called Black-Scholes model, where the volatility parameter σ plays a rather clear role, as we now discuss.

5 Black-Scholes Model (Single Stock)

The classical Black-Scholes model with constant coefficients contains the following two elements, for any $t \in [0, T]$ where T is a maturity or time horizon:

- A **risk-free account** B with constant rate r:

$$B(t) = e^{rt}.$$

- A **stock or index price process** S:

$$S(t) = S(0)e^{(\alpha - \frac{1}{2}\sigma^2)t + \sigma W(t)}.$$

Here, we use the nomenclature α = "mean rate of return" for stock S, and σ = "volatility" for stock S; while $\{W(t); t \geq 0\}$ is a **standard Brownian motion** (Wiener process).

The process W has the following properties:

- $W(0) = 0$, for $0 \leq s < t$, $W(t) - W(s)$ is independent of all the random variables $W(r)$ for $r \leq s$, and $W(t) - W(s)$ is centered normal with variance $t - s$.
- W has continuous paths with probability one.

- Convenient notation: $\mu := \alpha - \frac{1}{2}\sigma^2$.
- These parameters α, σ, μ are assimilated to quantities similarly to those in the discussion at the end of the previous section: specifically it holds that

$$\alpha = \lim_{h\downarrow 0} E[\frac{S(t+h)-S(t)}{h}]$$
$$\mu = \lim_{h\downarrow 0} E[\frac{1}{h}\log(\frac{S(t+h)}{S(t)})]$$
$$\sigma^2 = \lim_{h\downarrow 0} Var[\frac{S(t+h)-S(t)}{h}]$$
$$= \lim_{h\downarrow 0} Var[\frac{1}{h}\log(\frac{S(t+h)}{S(t)})].$$

Example 1 Today is time $t = T = Nh$; let $S_i = S(ih)$; $i = 0,1,\ldots,N$. Then σ^2 can be estimated as the rescaled empirical variance of the sequence of log returns:

$$\sigma^2 \approx \hat{\sigma}^2 \; : = \frac{1}{N}\sum_{i=0}^{N-1}\left(\frac{1}{\sqrt{h}}\log\left(\frac{S_{i+1}}{S_i}\right) - \frac{1}{\sqrt{h}}E\left[\log\left(\frac{S_{i+1}}{S_i}\right)\right]\right)^2$$
$$= \frac{1}{h}\frac{1}{N}\sum_{i=0}^{N-1}\left(\log\left(\frac{S_{i+1}}{S_i}\right) - \mu h\right)^2,$$

since $Var\left[\log\left(\frac{S_{i+1}}{S_i}\right)\right] = Var\left[\sigma W\left((i+1)h\right) - \sigma W\left(ih\right)\right] = \sigma^2 h$. Using same idea, we can also get estimators for μ and α.

5.1 *Method of Moments*

To explain the parameter choices made at the end of the previous section, one only needs to match the means and variances (first and second moments) of the stock returns over one period in the binomial model, to the same statistics in the Black-Scholes model in a period of length h, by using various specific choices for the objective probabilities of going up or down:

- We compute the log and simple returns in the Black-Scholes model:

$$\mathscr{R}_t = \log(\frac{S(t+h)}{S(t)}) = \mu h + \sigma(W(h+t) - W(t)),$$
$$R_t = \frac{S\left(t+h\right) - S\left(t\right)}{St} = e^{\mu h + \sigma(W(h+t)-W(t))} - 1,$$

so that we compute can compute their means and variances, and their asymptotics for small h:

$$\begin{aligned}\mathbf{E}[\mathscr{R}_t] &= \mu h,\\ Var[\mathscr{R}_t] &= \sigma^2 h,\\ \mathbf{E}[R_t] &= e^{\alpha h} - 1 \approx \alpha h,\\ Var[R_t] &= e^{(2\mu+\sigma^2)h}\left(e^{\sigma^2 h} - 1\right) \approx \sigma^2 h.\end{aligned}$$

- One possibility is to look for the binomial model with equal probabilities $p_u = 1 - p_u = 0.5$ of going up or down, and matching its simple returns' expectation and variance. Since then $R_t = (S_1 - S_0)/S_0 = u - 1$ or $d - 1$ with probabilities 0.5 and 0.5, those binomial statistics are

$$\begin{aligned}\mathbf{E}[R_t] &= \frac{u+d}{2} - 1,\\ Var[R_t] &= \frac{1}{2}\left((u-1)^2 + (d-1)^2\right) - \left(\frac{u+d}{2} - 1\right)^2\\ &= \left(\frac{u-d}{2}\right)^2.\end{aligned}$$

In the case of small h, this yields (approximately) the system

$$\begin{cases}\alpha h + 1 = \frac{u+d}{2}\\ \sigma\sqrt{h} = \frac{u-d}{2}\end{cases}$$

whose solution is easily seen to be

$$u = 1 + \alpha h + \sigma\sqrt{h}, d = 1 + \alpha h - \sigma\sqrt{h}$$

$$p_u = \frac{1}{2}.$$

We recognize the **Hull-White parametrization**.

- Another possibility is to decide that one prefers to have up and down factors which are reciprocals of each other. By inspecting the Black-Scholes model, ignoring the drift term μt and concentrating only on the term σW_t inside the exponential, one knows that an order of magnitude of the change of σW_t over a period of length h is its standard deviation, namely $\sigma\sqrt{h}$. It is then legitimate to require that $u = \frac{1}{d} = \exp(\sigma\sqrt{h})$. However, let us use the method of moments using only the restriction $u = 1/d$. We can compute mean and variance of the log return $\mathscr{R}_t$ in the one peroid binomial, finding

$$\begin{aligned}\mathbf{E}\left[\mathscr{R}_t\right] &= p_u \log u + (1-p_u)\log u^{-1}\\ &= (2p_u - 1)\log u\\ Var\left[\mathscr{R}_t\right] &= p_u \log^2 u + (1-p_u)^2 \log^2 u - (2p_u-1)^2 \log^2 u\\ &= \left(1 - (2p_{u-1})^2\right)\log^2 u.\end{aligned}$$

Using the approximation that if h is small, p should be close to $1/2$, matching the above variance with the Black-Scholes variance yields $\log^2 u = \sigma^2 h$, which is precisely

$$u = \frac{1}{d} = e^{\sigma\sqrt{h}}.$$

Plugging this into the equation for matching expectations $(2p_u - 1)\log u = \mu h$, we get, $\sigma\sqrt{h}\,(2p_u - 1) = \mu h$, i.e.

$$p_u = \frac{1}{2} + \frac{\mu}{2\sigma}\sqrt{h}.$$

This is the **Cox-Ross-Rubenstein** parametrization.

- One last possibility we examine is the case of matching means and variances of the log returns when $p_u = 1/2$. In this case, we compute those statistics for the Binomial model

$$\begin{aligned}\mathbf{E}\left[\mathscr{R}_t\right] &= \frac{1}{2}(\log u + \log d)\\ Var\left[\mathscr{R}_t\right] &= \frac{1}{2}\left(\log^2 u + \log^2 d\right) - \frac{1}{4}(\log u + \log d)^2\\ &= \left(\frac{\log u - \log d}{2}\right)^2.\end{aligned}$$

In this case, the moment-matching equations can be solved without resorting to approximations, and one finds

$$\begin{aligned}u &= e^{\mu h + \sigma\sqrt{h}},\\ d &= e^{\mu h - \sigma\sqrt{h}},\\ p_u &= \frac{1}{2}.\end{aligned}$$

This is the **Jarrow-Rudd** parametrization. This parametrization is closest in spirit to the original Black-Scholes model, if one attempts to discretize it in time by replacing each Brownian increment $W_{t+h} - W_t$ by a random variable taking the values $+\sigma\sqrt{h}$ and $-\sigma\sqrt{h}$ with equal probabilities, owing to the standard deviation and symmetry of the normal law for this increment. In fact, the binomal model with Jarrow-Rudd parameters converges to the Black-Scholes model. The proof of this fact is nearly immediate for fixed t by using the central limit theorem; that the convergence also holds at the process level (i.e. for all t simultaneously) is an application of the infinite-dimensional (functional) extension of the central limit theorem, sometimes known as Donsker's invariance principle.

The fact that there are several parametrization choices show that the binomal model is in fact richer than the Black-Scholes model; the former has one more parameter than the latter, hence the existence of many parametrization choices.

5.2 *Option Pricing Under BS Model*

Because of the close similarity between the binomial model with Jarrow-Rudd parameters and the Black-Scholes model, one suspects that the discounted risk-neutral valuation formula should hold for the Black-Scholes model. This is in fact true, and there is a generic option-pricing meta-theorem, which is broader than merely the Black-Scholes model, and also includes a statement about hedging.

Pricing metatheorem Let S be a stock price model, and let $\mathscr{X}$ be a contingent claim expiring at time T, i.e. $\mathscr{X}$ is a random variable which can be determined at time T using knowledge of the path of the stock price S up to time T. If the model for S can be expressed with a probability measure $\mathbf{Q}$ under which $t \to e^{-rt}S(t)$ is a martingale with respect S, then all contingent claims can be simultaneously priced in a consistent way via the formula

$$\text{price of } \mathscr{X}_T \text{ at time } 0 = e^{-rT}\mathbf{E}^{\mathbf{Q}}[\mathscr{X}_T].$$

If the measure $\mathbf{Q}$ is unique, then the price of every contingent claim is unique, and each such claim can be perfectly hedged (in continuous time) using a continuously-rebalanced self-financing portfolio of stock S and risk-free asset B.

In the case of Markovian models such as the Black-Scholes model, much more can be said about simple claims. We state the result in the Black-Scholes case only, for simplicity.

Definition 1 We say that $\mathscr{X}$ is a simple "contingent claim" (= a simple "option") if there as a non-random function $\Phi : \mathbb{R}_+ \to \mathbb{R}$ such that $\mathscr{X} = \Phi(S(T))$ (here T = maturity).

Theorem 1 (Discounted Risk-Neutral Valuation Formula) *Assume S satisfies the Black-Scholes model. The price P_t at time $t \leq T$ for the claim $\mathscr{X}$ defined above, is given by*

$$P_t = F(t, S(t))$$

where the non-random function $F : [0,T] \times \mathbb{R}_+ \to \mathbb{R}$ is given by

$$F(t,x) = \exp(-r(T-t))\mathbf{E}^*[\Phi(S(T))/S(t) = x],$$

where $\mathbf{E}^*$ *is the expectation under* $\mathbf{P}^*$ *the unique risk-neutral (martingale) measure. Moreover,* $\mathbf{P}^*$ *can be defined by saying that under* $\mathbf{P}^*$*, the parameter* α *in the Black-Scholes model need only be replaced by the risk-free rate* r.

5.3 Black-Scholes Formula

We may now use the meta-theorem's application to the Black-Scholes model, which we have just stated, to calculate the price of call options.

Definition 2 Let K be a positive constant. A simple contingent claim with contract function given by $\Phi(x) = \max(x - K, 0)$, is a Call option with strike price K.

We wish to compute $F(t,x)$ in the previous theorem when $\Phi(x) = \max(x - K, 0)$ and $S(t) =$ Black-Scholes model under $\mathbf{P}^*$. This can be done to a large extent by hand:

$$\begin{aligned} F(t,x) &= \exp(-r(T-t))\mathbf{E}^*\left[\max(x - K,0)/S(t) = x\right] \\ &= \exp(-r(T-t))\mathbf{E}[\max(x\exp(\mu^*(T-t) + \sigma(W(T) - W(t))) - K,0)] \\ &= \exp(-r(T-t))\mathbf{E}[\max(x\exp(\mu^*(T-t) + \sigma\sqrt{T-t}Z) - K,0)] \end{aligned}$$

where in the second line, we denote

$$\mu^* := r - \sigma^2/2$$

(i.e. we use the parameters for S under $\mathbf{P}^*$), and in the last line, thanks to scaling for normal laws, we assume that Z is a standard normal random variable. Note that, starting with the second displayed line above, it becomes unnecessary to add a star to the expectation sign. We also see that the last expression above depends on T and t only via $T - t$. Hence without loss of generality we set $t = 0$. Thus

$$\begin{aligned} F(0,x) &= e^{-rT}\mathbf{E}\left[\max(x\exp(\mu^*T + \sigma\sqrt{T}Z) - K,0)\right] \\ &= e^{-rT}\mathbf{E}\left[[\mathbf{1}_{\{x\exp(\mu^*T+\sigma\sqrt{T}Z)>K\}}(x\exp(\mu^*T + \sigma\sqrt{T}Z) - K)\right] \\ &= xe^{-rT}\mathbf{E}\left[\mathbf{1}_{\{x\exp(\mu^*T+\sigma\sqrt{T}Z)>K\}}\exp(\mu^*T + \sigma\sqrt{T}Z)\right] \\ &\quad -Ke^{-rT}\mathbf{P}\left[x\exp(\mu^*T + \sigma\sqrt{T}Z) > K\right]. \end{aligned}$$

We first compute the second piece:

$$\begin{aligned} P\left[x\exp(\mu^*T+\sigma\sqrt{T}Z)>K\right] &= P\left[Z>\frac{1}{\sigma\sqrt{T}}\left(-\mu^*T+\log\left(\frac{K}{x}\right)\right)\right] \\ &= P\left[Z<\frac{1}{\sigma\sqrt{T}}\left(\mu^*T+\log\left(\frac{x}{K}\right)\right)\right] \\ :&= P\left[Z<d_2\right] \\ :&= N(d_2) \end{aligned}$$

where N is the cummulative distribution function of the standard normal $N(d)=(2\pi)^{-1/2}\int_{-\infty}^{d}e^{-z^2/2}dz$ and

$$d_2:=\frac{1}{\sigma\sqrt{T}}\left(\mu^*T+\log\left(\frac{x}{K}\right)\right)=\frac{1}{\sigma\sqrt{T}}\left(\left(r-\sigma^2/2\right)T+\log\left(\frac{x}{K}\right)\right).$$

For the first piece, the computation is not much harder:

$$\begin{aligned} & xe^{-rT}\mathbf{E}\left[\mathbf{1}_{\{x\exp(\mu^*T+\sigma\sqrt{T}Z)>K\}}\exp(\mu^*T+\sigma\sqrt{T}Z)\right] \\ &= xE\left[\exp\left(-\frac{1}{2}\sigma^2T+\sigma\sqrt{T}Z\right)\mathbf{1}_{\{Z>-d_2\}}\right] \\ &= xE\left[\exp\left(-\frac{1}{2}\sigma^2T-\sigma\sqrt{T}(-Z)\right)\mathbf{1}_{\{-Z<d_2\}}\right] \\ &= xe^{-rT}\int_{-\infty}^{d_2}\exp\left(-\frac{1}{2}\sigma^2T-\sigma\sqrt{T}z\right)\exp(-\frac{z^2}{2})\frac{dz}{\sqrt{2\pi}} \\ &= x\int_{-\infty}^{d_2}\exp\left(-\frac{1}{2}(z+\sigma\sqrt{T})^2\right)\frac{dz}{\sqrt{2\pi}} \\ &= x\int_{-\infty}^{d_2+\sigma\sqrt{T}}\exp\left(-\frac{z^2}{2}\right)\frac{dz}{\sqrt{2\pi}} \\ &= xN(d_2+\sigma\sqrt{T}) \\ &:= xN(d_1) \end{aligned}$$

where

$$d_1:=\frac{1}{\sigma\sqrt{T}}\left(\left(r+\sigma^2/2\right)T+\log\left(\frac{x}{K}\right)\right).$$

We proved the famous Black-Scholes formula for pricing.

Theorem 2 *The pricing function F of the call option with strike K under the standard Black-Scholes model is:*

$$F(t,x)=xN(d_1)-Ke^{-r(T-t)}N(d_2)$$

where N is the standard normal distribution function and

$$
\begin{aligned}
d_1 &: = \frac{1}{\sigma\sqrt{T-t}}(\log(\frac{x}{K}) + (r + \frac{1}{2}\sigma^2)(T-t)),\\
d_2 &: = \frac{1}{\sigma\sqrt{T-t}}(\log(\frac{x}{K}) + \left(r - \frac{1}{2}\sigma^2\right)(T-t))\\
&= d_1 - \sigma\sqrt{T-t}.
\end{aligned}
$$

Remark 1 This formula has an interesting feature, which is to suggest a possible way of hedging the option over time: indeed, if at time t, where the stock price is $S(t)$, one invests in $N(d_1)$ shares of stock (here x must be replaced by the current stock value $S(t)$ in the formulas), then by also holding

$$F(t,S(t)) - S(t)N(d_1) = -Ke^{-r(T-t)}N(d_2)$$

dollars in the risk-free asset, one obtains a replicating portfolio for the option, i.e. one whose value is exactly that of the option at all times. This will be a worthwhile observation if one can prove that the portfolio is self-financing.

Remark 2 As it turns out, the previous portfolio really is self-financing, meaning that all changes in the portfolio allocations can be financed by the changes in the asset prices. We record this fact formally here, but rather enter into a formal proof, in the next section, we investigate how far from a perfect hedge one might get when the hedging portfolio is followed appriximately, by using discrete time.

Theorem 3 (Perfect Black-Scholes Hedge) *The Call option with strike-price K and maturity T can be perfectly replicated using the following portfolio: y_t shares of stock S and b_t dollars in the risk-free asset, with*

$$
\begin{aligned}
y_t &= N(d_1),\\
b_t &= -Ke^{-r(T-t)}N(d_2)
\end{aligned}
$$

where x is replaced by $S(t)$ in the formulas for d_1 and d_2. This portfolio is self-financing.

More generally, for the pricing function F of a given contingent claim $\mathcal{X} = \Phi(S(T))$, under the Black-Scholes model, the hedging portfolio defined by

$$y_t := \frac{\partial F}{\partial x}(t,S(t)); \quad b_t := F(t,S(t)) - S(t)\,y_t$$

is replicating by definition, and is self-financing.

6 Imperfect Black-Scholes Hedging in Discrete Time

6.1 Overnight Rebalancing, with Profit (Loss) Calculation

We mentioned earlier that following a trading strategy in continuous time is not practical. The replicating strategy (y_t, b_t) in the previous theorem is not immune to this difficulty. In practice, a high-frequency trading strategy (e.g. rebalancing every 5 min) can take advantage of rapid changs in stock values, but is too expensive to implement because of transaction costs.

Question: What happens if we follow the Black-Scholes hedging strategy only once a day?

We will use the arguments (t,x) and $(t+h, x+\varepsilon)$ for expressions F, d_1 and d_2, as shorthand notation, with the following understanding:

$$\begin{aligned} t &= \text{today} \\ h &= 1/365 \\ t+h &= \text{tomorrow} \\ x &= S(t) = \text{price of stock today} \\ x+\varepsilon &= S(t+h) = \text{price of stock tomorrow.} \end{aligned}$$

Taking the perspective of the option hedger, we sell one option at time t and purchase its corresponding Black-Scholes perfect hedging portfolio at the same time, and hold that portfolio without any rebalancing until time $t+h$. The value of the portfolio at time t is 0 by the hedging theorem. Let us investigate the so-called "overnight profit (or loss)"; this is thus identical to the value of the portfolio at time $t+h$ before rebalancing:

- Value held in option: $-F(t+h, x+\varepsilon)$.
- Value held in stock: $(x+\varepsilon) N(d_1(t,x))$.
- Value held in risk-free asset: $(F(t,x) - xN(d_1)) e^{rh}$.
- Total value held is: **Overnight profit (or loss)**

$$= -F(t+h, x+\varepsilon) + (x+\varepsilon) N(d_1(t,x)) + (F(t,x) - xN(d_1(t,x))) e^{rh}.$$

Example 2 In our numerical applications, we use: $x = 40$; $\varepsilon = 0.5$; $\sigma = 0.3$; $r = 0.08$; $h = 1/365$; we choose to price the call with $K = 40$ and $T - t = 1/4$ (3 months).

By using BS formula we find $F(t,x) = 2{,}7847$; $N(d_1(t,x)) = 0{,}5825$; $F(t,x) - xN(d_1) = -20{,}5159$; $F(t+h, x+\varepsilon) = 3{,}0665$. Thus in this case:

$$\begin{aligned} \textbf{Option hedger's overnight profit} &= -3{,}0665 + 0{,}5824 \times 40{,}5 - 20{,}5159 e^{0.08/365}. \\ &= 0{,}00500. \end{aligned}$$

This is great: this is very close to 0, so there is probably no need to buy or sell stock to rebalance the portfolio at time $t + h$.

Example 3 However, the good news in the previous example is due to the fact that the stock price did not move very far (only about 1 %) overnight. One might run into more trouble if the movements are larger. Repeating the previous calculations for various values of ε, we obtain the following Figures.

Figure 1 shows that when large stock price movements occur, the overnight profit quickly becomes a substantial loss. Figure 2 shows in detail the small magnitude of profit or loss when stock price changes are small (for $S(t + h)$ near $S(t)$.) An actual small profit occurs for $|S(t + h) - S(t)| < 0.6$ only.

We finish this section by mentioning the general form of the overnight profit under the Black-Scholes model.

Theorem 4 (Overnight profit for general simple claims under the Black-Scholes model) *With the pricing function F of a given contingent claim $\mathscr{X} = \Phi(S(T))$, under the Black-Scholes model, by following the hedging portfolio defined by $y_t := \frac{\partial F}{\partial x}(t,S(t))$; $b_t := F(t,S(t)) - S(t)\, y_t$ at discrete time intervals of length h, the overnight profit at time t is*

$$-F(t + h,x + \varepsilon) + (x + \varepsilon)\frac{\partial F}{\partial x}(t,x) + \left(F(t,x) - x\frac{\partial F}{\partial x}(t,x)\right)e^{rh}.$$

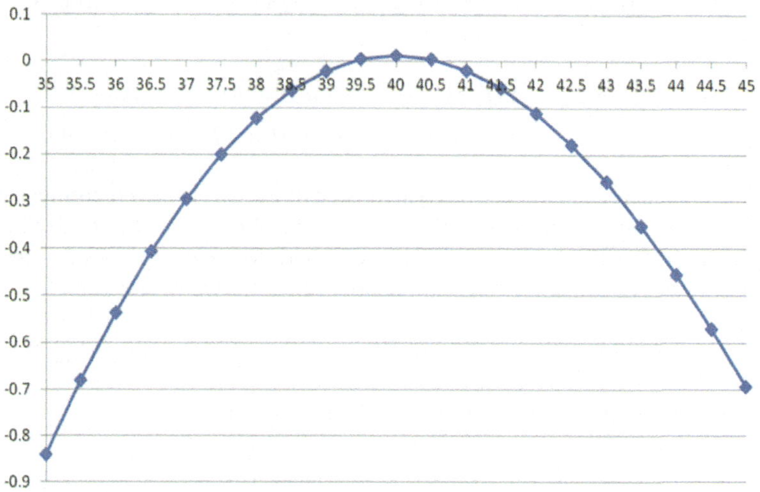

Fig. 1 Overnight profit for values of $x + \varepsilon$ from 35 to 45. Vertical axis is dollar value

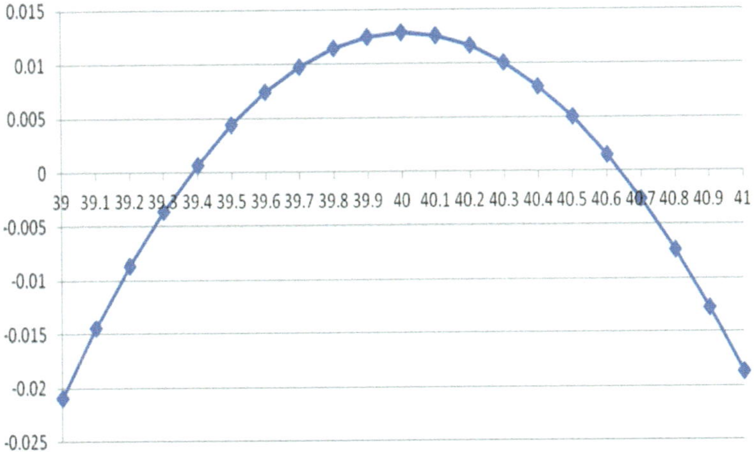

Fig. 2 Detail of overnight profit for values of $x + \varepsilon$ from 39 to 41

6.2 *Forecasting*

To understand the previous graph in theory, instead of analyzing the Black-Scholes formula in a mechanistic way (i.e. looking only at its dependence on t and x as a deterministic function), one may return to the probabilistic understanding and employ some simple forecasting. A rough approximation is in fact sufficient. Under the Black-Scholes model we can compute

$$\begin{aligned} S(t+h) - S(t) &= S(t) \exp(\mu h + \sigma(W(t+h) - W(t))) - S(t) \\ &= S(t)(\exp(\mu h + \sigma(W(t+h) - W(t))) - 1). \end{aligned}$$

Since h is considered small, and the typical size of the mean-zero increment $W(t+h) - W(t)$ is the size of its standard deviation, i.e. $\sqrt{h}$, one may consider in a first approximation that $W(t+h) - W(t)$ is small and dominates μh. Thus, using the first order Taylor expansion of the exponential function, we would get

$$S(t+h) - S(t) = S(t)(\sigma(W(t+h) - W(t))) + o(\sqrt{h}).$$

We may interpret this approximation in a binary way, as

$$S(t+h) - S(t) \simeq (\pm 1)\sigma\sqrt{h}S(t).$$

In other words, with the x and ε notation, this approximation is equivalent to:

$$\varepsilon \simeq (\pm 1)\sigma\sqrt{h}x$$

where the (± 1) symbol represents a random variable which takes values $+1$ and -1 with equal probabilities.

Using Taylor's formula on F up to order 1 in time and order 2 in space, we obtain

$$F(t+h,x+\varepsilon) = F(t,x) + h\frac{\partial F}{\partial t}(t,x) + \varepsilon\frac{\partial F}{\partial x}(t,x) + \frac{1}{2}\varepsilon^2\frac{\partial^2 F}{\partial x^2}(t,x) + o(h) + o(\varepsilon^2).$$

Therefore,

$$\begin{aligned}\text{overnight profit} = &-\left[F(t,x) + h\frac{\partial F}{\partial t}(t,x) + \varepsilon\frac{\partial F}{\partial x}(t,x) + \frac{1}{2}\varepsilon^2\frac{\partial^2 F}{\partial x^2}(t,x)\right]\\ &+\frac{\partial F}{\partial x}(t,x)\,(x+\varepsilon) + \left[F(t,x) - x\frac{\partial F}{\partial x}(t,x)\right](1+rh) + o(h) + o(\varepsilon^2).\end{aligned}$$

One notes that all the terms involving ε rather ε^2 miraculously disappear, as do all the terms which are not small: this is because the Black-Scholes hedge was chosen in such a way to make these simplications occur, at least in the first approximation we are using here. Thus we get *overnight profit*

$$= -h\frac{\partial F}{\partial t}(t,x) - \frac{1}{2}\varepsilon^2\frac{\partial^2 F}{\partial x^2}(t,x) + rhx\frac{\partial F}{\partial x}(t,x) + rhF(t,x) + o(h) + o(\varepsilon^2).$$

Interestingly, in a first-order approximation on ε, one sees that if $\frac{\partial^2 F}{\partial x^2}(t,x) > 0$, which is typically the case for most options, the highest overnight profit is obtained when ε is 0. Now using the forecast for ε, we see that $\varepsilon^2 = \sigma^2 h$ and that $o(h) = o(\varepsilon^2)$. This yields

$$\text{overnight profit} = \left(rF(t,x) - \frac{\partial F}{\partial t}(t,x) - \frac{1}{2}\sigma^2 x^2\frac{\partial^2 F}{\partial x^2}(t,x) - rx\frac{\partial F}{\partial x}(t,x)\right)h + o(h).$$

The option hedger must try to keep her profit to a minimum in absolute value, as one should from the perspective of an insurer, which is to minimize risk (it is also a good idea from an investment perspective, since we saw in the previous section that the overnight profit's downside is significantly greater than its upside.) This risk-minimizing strategy can thus be summarized as "Overnight profit = 0"

$$\Longleftrightarrow \quad \frac{\partial F}{\partial t}(t,x) + rx\frac{\partial F}{\partial x}(t,x) + \frac{1}{2}\sigma^2 x^2\frac{\partial^2 F}{\partial x^2}(t,x) - rF(t,x) = 0.$$

This is precisely the famous so-called **BLACK-SCHOLES PDE** !

What we have essentially just shown is that the Black-Scholes hedge is a perfect self-financing replicating portfolio if and only if the Black-Scholes PDE holds for the pricing function F. In fact, the above development is in some sense equivalent to the classical proof of the Black-Scholes pricing and hedging theorem by means of stochastic calculus and the Itô formula. We do not provide the details, but state the result formally, also summarizing our theorems on the discounted risk-neutral valuation formula and the perfect Black-Scholes hedge.

Theorem 5 *For a generic simple European contingent claim $\mathscr{X} = \Phi(S(T))$ under the Black-Scholes model, let*

$$F(t,x) = e^{-r(T-t)}\mathbf{E}^*\left[\mathscr{X} \mid S(t) = x\right]$$

where under $\mathbf{P}^*$, the parameter α is replaced by the short rate r. Then $\mathscr{X}$ can be replicated perfectly at all times $t \leq T$ by using the following Black-Scholes self-financing replicating portfolio:

1. *At time $t_0 < T$, collect $F(t_0, S(t_0))$ dollars from the sale of $\mathscr{X}$.*
2. *At all times $t \in [t_0, T)$, compute $\frac{\partial F}{\partial x}(t,x)$, and invest (long or short) in $y_t = \frac{\partial F}{\partial x}(t, S(t))$ shares of stock at time t.*
3. *At the same time, put (or borrow) $b_t = F(t, S(t)) - y_t S(t)$ dollars in the risk-free account to finance the stock investment.*

This strategy is possible because the portfolio is self-financing. Moreover, F solves the Black-Scholes PDE with terminal condition $F(t,x) = \Phi(x)$ for all $x > 0$ and all $t \in [t_0, T]$.

6.3 Delta and Gamma Hedging

To improve on the discrete hedging strategy studied above, also known as the discrete Delta hedge (recall from the previous graphs that the overnight losses can be substantial if there are large price changes) we consider a possible second-order approximation to the perfect Black-Scholes hedge.

Definition 3 For any $\mathscr{X} = \Phi(S(T))$ with pricing function F let

$$\Delta_F(t,x) = \text{“Delta”} = \frac{\partial F}{\partial x}(t,x),$$

$$\Gamma_F(t,x) = \text{“Gamma”} = \frac{\partial^2 F}{\partial x^2}(t,x).$$

These are two examples of what we call "Greeks", sensitivities of pricing functions to changes in their parameters. Other greeks include the Theta $\Theta = \frac{\partial F}{\partial(T-t)}$, the Rho $\rho = \frac{\partial F}{\partial r}$, and the Vega $\mathscr{V} = \frac{\partial F}{\partial \sigma}$ (even though Vega is not really a Greek letter!!).

Similarly to what is done in the insurance business, we can look for a way of transferring some of the risk in the Delta-hedging portfolio to a third party, i.e. a reinsurance contrat. We show how this works on an example.

Example 4 Back to our call with $K = 40$ and $T = 1/4$, imagine that we worry that $S(t)$ will go much higher than 40 overnight. We will ask another market maker to sell us a call option with $K' = 45$, and a longer expiration $T' = 1/3$ (4 months).

- **Whole portfolio**: our new portfolio has -1 unit of $K = 40$-call (pricing function F_{40}), y_t shares of S, z_t units of the $K' = 45$-call (pricing function F_{45}), and b_t in risk-free account which we compute to make the value at time 0 of our entire portfolio equal to 0. Its value is

$$0 = V(t,x) := -F_{40}(t,x) + y_t x + z_t F_{45}(t,x) + b_t.$$

- **Goal**: for the whole portfolio with value $V(t,x)$, we want not just $\Delta_V = \frac{\partial V}{\partial x} = 0$ but also $\Gamma_V = \frac{\partial^2 V}{\partial x^2} = 0$. In the old portfolio we just had $\frac{\partial V_{\text{old}}}{\partial x} = 0$.
- **Gamma condition**. Slightly abusively, we consider that partial derivatives operate only on pricing functions (this is an excellent approximation, it turns out):

$$\Gamma_V = \frac{\partial^2 V}{\partial x^2} = -\Gamma_{40}(t,x) + y_t \times 0 + z_t \Gamma_{45}(t,x).$$

  Since we want $\Gamma_V = 0$, this yields the choice

$$z_t = \frac{\Gamma_{40}(t,x)}{\Gamma_{45}(t,x)}.$$

- **Delta condition**. Next, with z_t already computed, we calculate

$$\Delta_V = -\Delta_{40}(t,x) + y_t + z_t \Delta_{45}(t,x),$$

  and wanting $\Delta_V = 0$, this gives

$$y_t = \Delta_{40}(t,x) - z_t \Delta_{45}(t,x).$$

- **Cash.** Finally, since y_t and z_t have been computed, we now compute the risk-free position:

$$b_t = F_{40}(t,x) - y_t x - z_t F_{45}(t,x).$$

Remark 3 We already know that for the call F_K we have $\Delta_K(t,x) = N(d_1(t,x))$. Therefore, by the chain rule, since $N'(z) = (2\pi)^{-1/2} e^{-z^2/2}$, and

$$\partial d_1/\partial x = 1/\left(\sigma x \sqrt{T-t}\right),$$

we get

$$\Gamma_K(t,x) = \frac{1}{\sigma x \sqrt{T-t}\sqrt{2\pi}} e^{-d_1(t,x)^2/2}.$$

Example 5 With the same parameters as previously $r = 0.08$, $\sigma = 0.3$, $K = 40$, $K' = 45$, $T - t = 1/4$, $T' - t = 1/3$, using the above formulas, we can compute

at t	40-call	45-call
F	2.7847	1.3741
Δ	0.5825	0.3301
Γ	0.06794	0.06342

from which the expressions for the allocations of stock and 45-call become

$$z_t = \frac{0.06794}{0.06342} = 1.0714,$$
$$y_t = 0.5825 - 1.0714 * 0.3301 = 0.2288.$$

By repeating the overnight profit analysis here we find

$$\textbf{Overnight profit (or loss)}$$
$$= -F_{40}(t+h,x+\varepsilon) + (x+\varepsilon)y_t + F_{45}(t+h,x+\varepsilon)z_t$$
$$+ (F_{40}(t,x) - xy_t - F_{45}(t,x)z_t)e^{rh}.$$

Thus if $\varepsilon = 0.5$ for instance, one finds $F_{40}(t+h,40.5) = 2.767$ and $F_{45}(t+h, 40.5) = 1.361$, so that the overnight profit computes to 0.001813, which is about one third of what it was for the pure discrete Delta-hedging strategy. This decrease

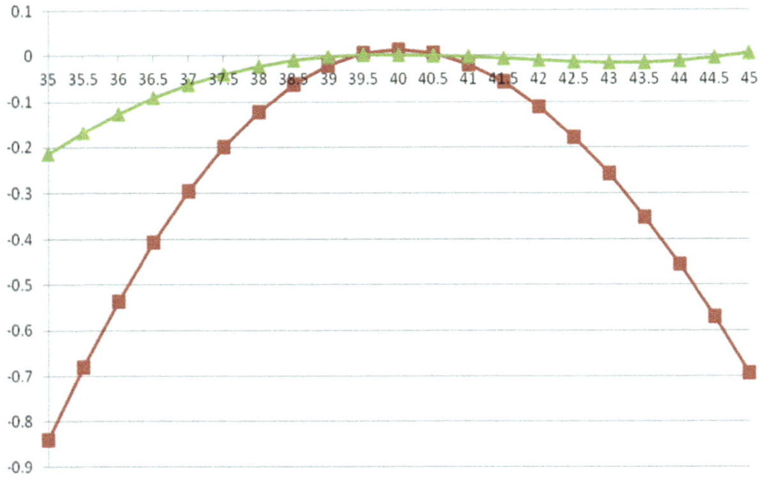

Fig. 3 Overnight profit for values of $x + +\varepsilon$ from 35 to 45: *green line* is Delta hedge, *red line* is Delta and Gamma hedge using a 45-strike 4-month call. From 39 to 41

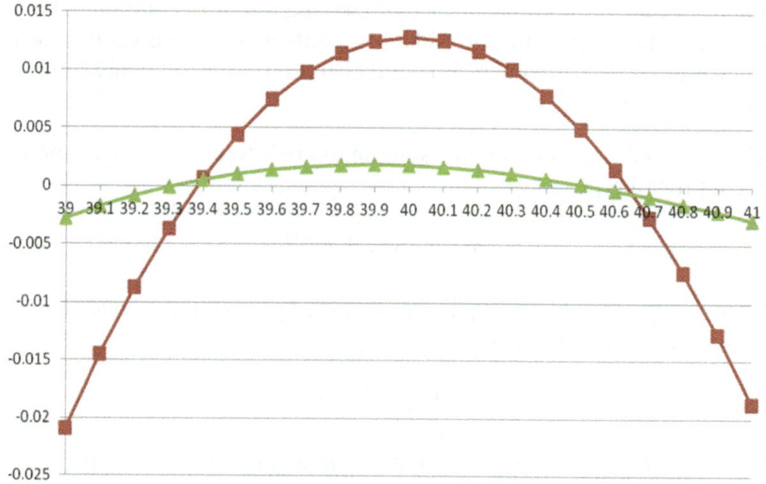

Fig. 4 Overnight profit for values of $x+\varepsilon$ from 35 to 45: *green line* is Delta hedge, *red line* is Delta and Gamma hedge using a 45-strike 4-month call

denotes better risk-management. The improvement is particularly evident for large values of ε, as can be seen in Figs. 3 and 4.

The next graph shows the detail for small stock movements: here too, the improvement of the Delta-Gamma hedge over the Delta hedge is marked.

7 Extensions of the Black-Scholes Formula

Recall that in a Call option, the strike is denoted by K, and the stock by S, but in reality, by introducing the concept of prepaid forward prices for assets, these notions become relative, and may be switched for convenience. An example of such a situation is that where K is a second asset, and the call option is then an exchange option. Let us be more precise about the generic framework.

- Main Idea: the prepaid forward price of any quantity K over $[t, T]$ is the cash value needed on hand at time t to guarantee a payoff of K at time T. This would hold whether K is non-random, or a traded risky asset such as a stock or an interest rate or an index, or even if it is a contingent claim. In the last case, the prepaid forward price is what we have simply been calling the price of the claim.
- Let us discuss the other cases. Generally, we may use the notation $F^P_{t,T}$ for the operator which computes the prepaid forward prices of assets. We assume for

simplicity, as we have before, that the risk-free rate r is constant. The general principle by which it is sufficient to identify a self-financing replicating portfolio to compute prepaid forward prices still holds. This makes the following computations essentially trivial.

- When K is non-random, by investing in the risk-free asset alone, one finds its prepaid forward price as

$$F_{t,T}^{P}(K) = Ke^{-r(T-t)}.$$

- When K is a non-dividend-paying stock S, by investing in one share of this stock alone, by definition, one finds

$$F_{t,T}^{P}(S) = S(t).$$

- When a stock pays a continuous dividend rate δ, this means that by purchasing the stock at the price $S(t)$ at time t, one will obtain at time T the value $S(T)\, e^{\delta(T-t)}$. Therefore, by investing in a discounted number of shares of this stock alone, we find

$$F_{t,T}^{P}(S) = S(t)e^{-\delta(T-t)}.$$

- If S pays a discrete dividend of \$ D at a fixed time $u \in [t,T]$, this means that an investment in one share worth $S(t)$ at time t yields one share worth $S(T)$ at time T but also a fixed payoff of D dollars at time $u < T$. Thus an investment in one share of stock minus a discounted amount borrowed at the risk free rate from time t to time u, will replicate the stock's payment stream, yielding

$$F_{t,T}^{P}(S) = S(t) - De^{-r(u-t)}.$$

- Combining the above two dividends, we get in general

$$F_{t,T}^{P}(S) = S(t)e^{-\delta(T-t)} - De^{-r(u-t)}.$$

In the formulas below, as usual, the $S(t)$ is to be replaced by x.

- It turns out that we can recast the classical BS formula for calls with no dividends and $K =$ constant in terms of prepaid forward prices for S and K: using the notation x instead of $S(t)$, as usual, we have

$$\begin{aligned}\text{price of the call } C(t,x) &= xN(d_1) - Ke^{-r(T-t)}N(d_2)\\ &= F_{t,T}^{P}(S)N(d_1) - F_{t,T}^{P}(K)N(d_2).\end{aligned}$$

$$d_2 = \frac{1}{\sigma\sqrt{T-t}}\left(\log\left(\frac{x}{K}\right) + \left(r - \frac{1}{2}\sigma^2\right)(T-t)\right)$$
$$= \frac{1}{\sigma\sqrt{T-t}}\left(\log\left(\frac{F^P_{t,T}(S)}{F^P_{t,T}(K)}\right) - \frac{1}{2}\sigma^2(T-t)\right);$$
$$d_1 = d_2 + \sigma\sqrt{T-t}.$$

This formula extends in many cases.

- **Options for dividend-paying stocks**

Theorem 6 *If S pay a single discrete dividend D at time $u < T$ and/or continuous dividends at the rate δ, the formula*

$$C(t,x) = F^P_{t,T}(S)N(d_1) - F^P_{t,T}(K)N(d_2)$$

for the price of the call holds true for all $t < u$ with d_1 and d_2 as above, and $F^P_{t,T}(K) = Ke^{-r(T-t)}$ and $F^P_{t,T}(S) = xe^{-\delta(T-t)} - De^{-r(u-t)}$.

Remark 4 When $D = 0$, more generally for a simple European claim $\mathscr{X} = \Phi(S(T))$ with contract function Φ, the pricing function F_δ for this claim satisfies the following modified Black-Scholes PDE with continuous dividend rate δ and terminal condition Φ

$$\frac{\partial F_\delta}{\partial t} + (r - \delta - \frac{1}{2}\sigma^2)\frac{\partial F_\delta}{\partial x} + \frac{1}{2}\sigma^2\frac{\partial^2 F_\delta}{\partial x^2} - rF_\delta = 0.$$

Remark 5 The solution of this PDE, and the hedging portfolio, are easily computed given our previous work. In fact, the following results hold.

Theorem 7 (Pricing and hedging with continuous dividends).

Remark 6

- The discounted risk-neutral valuation formula still holds, with $\mathbf{P}^*$ defined by replacing α by $r - \delta$ in the Black-Scholes model.
- Let F_0 be the solution to the Black-Scholes PDE with $\delta = 0$. Then the general solution is given by
$$F_\delta(t,x) = F_0\left(t, xe^{-\delta(T-t)}\right).$$
- The hedging portfolio for claim $\mathscr{X} = \Phi(S(T))$ is still given by investing in the following number of shares of stock:
$$y_t = \Delta(t,S(t)) := \frac{\partial F_\delta}{\partial x}(t,S(t)),$$
and financing this position by holding $b_t = F_\delta(t,S(t)) - S(t)y_t$ money in the risk-free account.

Remark 7 Note that, from the previous formula for F_δ via F_0, we have

$$\begin{aligned}\Delta(t,x) &= e^{-\delta(T-t)}\frac{\partial F_0}{\partial x}\left(t, xe^{-\delta(T-t)}\right)\\ &= e^{-\delta(T-t)}\Delta_0\left(t, xe^{-\delta(T-t)}\right)\end{aligned}$$

where Δ_0 is the stock hedging position with zero dividend. Hence to hedge an option on a continuous dividend-paying stock, one may use the zero-dividend hedge by reducing the position by the dividend discount factor, and reducing the current observed stock price by the same factor.

- **Option on currency exchange rate**
  Let $X(t)$ = exchange rate (e.g. the price in US dollars (domestic currency) for one Euro (foreign currency)). We use the Black-Sholes model with a volatility σ for X. Also, we have two risk-free rates to consider:

$$\begin{aligned}\text{Forigne risk - free rate } &= r_f,\\ \text{Domestic risk - free rate } &= r.\end{aligned}$$

  Since the Euro can be considered as a risky asset, and can also be invested in the foreign risk-free account, it will yield payments at a continuously compouded rate r_f when placed in this account. Therefore, X is just like a stock with continuous dividend rate $\delta = r_f$. This proves that a call option on X with strike price K has pricing function $F_\delta = F_{r_f}$.
- **Call option on a futures contrat**
  A Futures on a stock S in the interval $[t, T]$ is a contact in which the counterparties decide on price to pay for S at time t, but the stock is delivered at time T and the price is paid at delivery. We use a Black-Scholes model for the futures price $G(t)$. However, the prepaid forward price of a futures on S is not $G(t)$ but $e^{-r(T-t)}G(t)$, since an investment of that many dollars in the risk-free accout will yield the quantity $G(t)$ at time T, which is precisely the price to be paid for the stock S at time T under the futures contract. Hence we have

$$F_{t,T}^P(G) = e^{-r(T-t)}G(t).$$

  By comparing with the prepaid forward price of a dividend-paying stock, one sees that the price G of the futures contract on S is like the price of a version of S which pays a continuous dividend rate of $\delta = r$, and the corresponding pricing function $F = F_r$ for options. For instance, for the call option, one obtains a particularly simple pricing function

$$C_G(t,x) = xe^{-r(T-t)}N(d_1) - Ke^{-r(T-t)}N(d_2),$$
$$d_2(t,x) = \frac{1}{\sigma\sqrt{T-t}}\left(\log\left(\frac{x}{K}\right) - \frac{1}{2}\sigma^2(T-t)\right),$$
$$d_1(t,x) = d_2(t,x) + \sigma\sqrt{T-t}.$$

- **Exchange option**
Instead of using a constant strike K, let us use another stock $\tilde{S}$. Hence, we consider the contingent claim
$$\mathscr{X} = \max\left(S(T) - \tilde{S}(T), 0\right).$$

Specifically because this option is similar to a call, the general pricing principe still works, but one must reinterpret the volatility. By using an argument by which one rewrite the claim's payoff as $\mathscr{X} = \tilde{S}(T)\max\left(S(T)/\tilde{S}(T) - 1, 0\right)$, one realizes that the normalized asset $S(T)/\tilde{S}(T)$ plays an important role. Assuming that both assets satisfy Black-Scholes models with correlated Brownian motions, they have volatilities σ and $\tilde{\sigma}$, and one may assume that their Brownian motions W and $\tilde{W}$ have correlation ρ. This implies that the normalized asset $S/\tilde{S}$ has volatility equal to

$$\sigma_e := \sqrt{E\left[\left(\sigma W(1) - \tilde{\sigma}\tilde{W}(1)\right)^2\right]} = \sqrt{\sigma^2 + \tilde{\sigma}^2 - 2\rho\sigma\tilde{\sigma}}.$$

A further argument leads to realizing that for the normalized asset $S/\tilde{S}$, under a risk-neutral measure, the mean rate of return parameter α should be zero. This all leads to a pricing formula for $\mathscr{X}$ which follows the general call formula with $r = 0; \delta = 0$, and $\sigma = \sigma_e$ given above. Thus, with the notation x representing $S(t)$ and y representing $\tilde{S}(t)$, we get the exchange option pricing function

$$C_e(t,x,y) = xN(d_1) - yN(d_2),$$
$$d_2(t,x) = \frac{1}{\sigma_e\sqrt{T-t}}\left(\log\left(\frac{x}{y}\right) - \frac{1}{2}\sigma_e^2(T-t)\right),$$
$$d_1(t,x) = d_2(t,x) + \sigma_e\sqrt{T-t}.$$

When S and/or $\tilde{S}$ pay dividends, we get the usual modifications to the prepaid forward prices of x and y. More generally, we have the following call pricing general principle.

- **Conclusion: call pricing meta-theorem**
Let S and $\tilde{S}$ be two assets which could be dividend-paying, or not, or could be constants. Let σ be the effective volatility of the normalized asset $S/\tilde{S}$. Then the price of the contingent claim $\mathscr{X} = \max\left(S(T) - \tilde{S}(T), 0\right)$ is $C\left(t, S(0), \tilde{S}(0)\right)$ where

$$C(t) = F_{t,T}^{P}(S)\, N(d_1) - F_{t,T}^{P}\left(\tilde{S}\right) N(d_2),$$

$$d_2(t,x) = \frac{1}{\sigma\sqrt{T-t}}\left(\log\left(\frac{F_{t,T}^{P}(S)}{F_{t,T}^{P}\left(\tilde{S}\right)}\right) - \frac{1}{2}\sigma^2(T-t)\right),$$

$$d_1(t,x) = d_2(t,x) + \sigma\sqrt{T-t}.$$

8 An Even More General Black-Scholes Formula: Stochastic Interest Rates

One problem with the Black-Scholes model is that the parameters are assumed to be constant. This can be a problematic assumption over long time periods, particularly for the purpose of option pricing, where the time scale is in months. While arguably the biggest issue is the non-constance of the volatility parameter, here we will discuss the case where the short rate parameter is itself assumed to be a stochastic process. In this case, one cannot simply write $F_{t,T}^{P}(K) = e^{-r(T-t)}K$, but it turns out that the discounted risk-neutral valuation formula holds. Without entering into the details of determining bond models and prices, we state the following.

Definition 4 The zero-coupon bond with maturity T is a contract that yields one dollar at time T. Its price at time $t < T$ is denoted by $P(t,T)$.

Theorem 8 *Assume the short rate $r(t)$ is a stochastic process, and that there exists a martingale measure $\mathbf{Q}$ for the bond market. Then*

$$P(t,T) = \mathbf{E}^{\mathbf{Q}}\left[\exp\left(-\int_t^T r(s)\,ds\right)\right].$$

The general call-pricing principle described at the end of the previous section can be updated for this situation in which interest rates are stochastic, to some extent, using the idea of change of numeraire. We provide some ideas and a special case where the computations can be carried out.

Let S be a fixed asset. For any other asset $\tilde{S}$, we say that $\tilde{S}/S$ is the asset $\tilde{S}$ under the change of numeraire S. The probability measure $\mathbf{P}^S$, if it exists, is one under which each asset $\tilde{S}/S$ is a martingale; this $\mathbf{P}^S$ is called the "S-neutral measure". When one uses the bond prices $P(\cdot,T)$ as the normalizing asset, the measure $\mathbf{P}^{P(\cdot,T)}$ is usually denoted by $\mathbf{P}^T$, and is called the "T-Forward neutral measure".

Theorem 9 *For the standard call on S with strike K under stochastic interest rates given by a bond-price model $P(\cdot,T)$, the call pricing function is*

$$C(t,x) = x\mathbf{P}^S[S(t) \geq K \mid S(t) = x] - KP(t,T)P^T[S(t) \geq K \mid S(t) = x].$$

The problem with the above theorem is that computing the law of $S(t)$ under P^T and under P^S is sometimes difficult. The following special situation allows a full computation.

Example 6 **(El Karoui, Geman, Rochet)** Let

$$Z(t) = \frac{S(t)}{P(t,T)}.$$

The dynamics of Z may be highly non-trivial in terms of the mean rate of return of Z, but its volatility structure is the same under the measures $\mathbf{P}^T$ and $\mathbf{P}^S$ and the original objective probability measure $\mathbf{P}$. If this volatility happens to be time-dependent by non-random, i.e. equal to the function $\sigma(s)$ for $s \in [t,T]$, then $C(t,x)$ satisfies the standard Black-Scholes formula with effective volatilty

$$\sigma := \sqrt{\frac{1}{T-t}\int_t^T |\sigma(s)|^2 \, ds}.$$

9 For Further Analyses: Basic Introduction to the Black-Scholes Model with Itô calculus

9.1 *Itô calculus*

All the formulas developed for continuous-time models are typically shown rigorously by using Itô's stochastic calculus, which we now introduce briefly.

Let W = a brownian motion.

Let $f \in \mathscr{C}_b^2$ (twice continuously differentiable function with bounded derivatives), let $Y(t) = f(W(t))$. We may use Taylor's formula to express changes in the process Y:

$$\begin{aligned} Y(t+h) - Y(t) &= f'(W(t))(W(t+h) - W(t)) \\ &\quad + \frac{1}{2} f''((W(t))(W(t+h) - W(t))^2 + o\left((W(t+h) - W(t))^2\right). \end{aligned}$$

Let us investigate what happens when $h = dt$ is an infinitesimal.

Itô's rule Using standard Gaussian calculation rules, since $W(t+h) - W(t)$ is normal with mean zero and variance h, we find

$$E\left[(W(t+h) - W(t))^2\right] = h$$

and

$$
\begin{aligned}
Var\left((W(t+h)-W(t))^2\right) &= E\left[(W(t+h)-W(t))^4\right]-\left(E\left[(W(t+h)-W(t))^2\right]\right)^2\\
&= 3h^2-h^2=2h^2.
\end{aligned}
$$

Therefore with $h = dt$, ignoring terms of order dt^2, the random variable $(W(t+dt) - W(t))^2$ may be interpreted as one which has a zero variance. With the Itô differential notation

$$dW(t) = W(t+dt) - W(t)$$

this leads to the following:

Itô's rule: $(W(t+dt) - W(t))^2 = dt.$

Other Itô rules Similarly we obtain

$$dW(t) \cdot dt = 0$$

because $h(W(t+dt) - W(t)) = O(h)$. Let $\widetilde{W}$ be an independent copy of W : then

$$dW(t) \cdot d\widetilde{W}(t) = 0.$$

Itô's formula Using these rules in the earlier Taylor expansion, adding an extra time parameter for convenience, and integrating over time, we finally arrive at the following.

Theorem 10 (Itô's formula and Itô integral) *For $F(t,x)$ of class $C^{1,2}$*

$$
\begin{aligned}
f(t,W(t)) &= f(0,0) + \int_0^t \frac{\partial f}{\partial t}(u,W(u))\,du + \int_0^t \frac{\partial f}{\partial x}(u,W(u))\,dW(u)\\
&+\frac{1}{2}\int_0^t \frac{\partial^2 f}{\partial x^2}(u,W(u))\,du
\end{aligned}
$$

where the first and third integrals are of Riemann type, and the second is of the so-called Itô type, which can be defined as the limit in $L^2(\Omega)$ of its Itô-type Riemann-Stieltjes sums

$$\int_0^t g(W(s))\,dW(s) = \lim_{n\to\infty}\sum_{k=0}^{n-1} g\left(W\left(\frac{kt}{n}\right)\right)\left(W\left(\frac{(k+1)t}{n}\right) - W\left(\frac{kt}{n}\right)\right).$$

Itô's formula for the Black-Scholes model Let S satisfy the Black-Scholes model

$$S(t) = S_0 e^{\mu t + \sigma W(t)},\ \mu = \alpha - \frac{1}{2}\sigma^2.$$

The Itô formula above implies that S solves the Black-Scholes stochastic differential equation

$$dS(t) := S(t+dt) - S(t) = \alpha S(t)dt + \sigma S(t)dW(t).$$

The following Itô rule for S is a consequence of using Itô's rules formally on $(dS(t))^2$ above:

$$(dS(t))^2 = \sigma^2 S(t)^2 dt.$$

Indeed $(dS(t))^2 = \alpha^2(S(t))^2(dt)^2 + 2\alpha\sigma S^2(t)dtdW(t) + \sigma^2 S^2(t)(dW)^2$ and the first and second terms are 0. This can also be established by using the methodology employed for W directly for S itself. The Itô rule for S and Taylor's formula similarly imply the following:

Theorem 11 (Itô's formula for the Black-Scholes model) *Let S be a Black-Scholes model and $f \in C^{1,2}$, then for every $t \geq 0$,*

$$\begin{aligned} f(t,S(t)) &= f(0,S(0)) + \int_0^t \frac{\partial f}{\partial t}(u,S(u))\,du \\ &+ \int_0^t \frac{\partial f}{\partial x}(u,S(u))\,dS(u) + \frac{1}{2}\int_0^t \frac{\partial^2 f}{\partial x^2}(u,S(u))\,(dS(u))^2 \\ &= f(0,S(0)) + \int_0^t \frac{\partial f}{\partial t}(u,S(u))\,du \\ &+ \int_0^t \frac{\partial f}{\partial x}(u,S(u))\,(\alpha S(u)\,du + \sigma S(u)\,dW(u)) \\ &+ \frac{1}{2}\int_0^t \frac{\partial^2 f}{\partial x^2}(u,S(u))\,\sigma^2 S(u)^2\,du. \end{aligned}$$

Itô's rule for pairs of processes For specific models, one may use Itô's rules formally to evaluate the last term in the following general "integration by parts" principle: for X, Y two process:

$$d(X(t)Y(t)) = X(t)dY(t) + Y(t)dX(t) + dX(t)\cdot dY(t).$$

9.2 *Application: Self-financing Condition and Option Pricing and Hedging Theorem*

Consider $\{S_i; i = 1, \ldots, d\}$ a set of d risky Black-Scholes-type assets. The value of a portfolio which contains $y_i(t)$ shares of S_i at time t is

$$V(t) = \sum_{i=1}^{d} y_i(t) S_i(t).$$

By Itô's integration-by-parts formula

$$dV(t) = \sum_{i=1}^{d} y_i(t) dS_i(t) + dy_i(t) S_i(t) + dy_i(t) dS_i(t).$$

Let us investigate an infinitesimal interpretation of the notion of self-financing. In discrete time, when passing from time $t - h$ to time t, changes in portfolio positions must be financed by current changes in asset values. This means that the portfolio value just before changing allocations at time t must equal the portfolio value right after changing allocations at the same time t:

$$\sum_{i=1}^{d} S_i(t) h_i(t) = \sum_{i=1}^{d} S_i(t) h_i(t - h).$$

Consequently

$$\begin{aligned} 0 &= \sum_{i=1}^{d} S_i(t)(h_i(t) - h_i(t-h)) \\ &= \sum_{i=1}^{d} S_i(t-h)(h_i(t) - h_i(t-h)) - \sum_{i=1}^{d} (S_i(t) - S_i(t-h))(h_i(t) - h_i(t-h)). \end{aligned}$$

Then passing to $h = dt$ infinitesimal, we find

$$0 = \sum_{i=1}^{d} S_i(t) dh_i(t) - \sum_{i=1}^{d} dS_i(t) dh_i(t).$$

Combining this with the expression above for $dV(t)$, we arrive at the

$$\textbf{Self-financing condition: } dV(t) = \sum_{i=1}^{d} y_i(t) dS_i(t).$$

Note that the risk-free asset can be denoted by $S_0(t)$ and satisfies $dS_0(t) = rS_0(t)\,dt$. Thus if instead of denoting by $y_0(t)$ the number of "shares" of the risk-free asset, we use the notation b_t for the cash amount in the risk-free asset, then $b_t = y_0(t) S_0(t)$ and thus $y_0(t) dS_0(t) = b_t r dt$. Consequently we have

Self-financing condition for a portfolio of y_t shares of stock and b_t in the risk-free asset:

$$dV(t) = y_t dS(t) + b_t r dt.$$

It is now possible to check that the main pricing and hedging theorem for the Black-Scholes model is correct. Recall that it states that with F satisfying the Black-Scholes PDE with terminal condition Φ, the claim $\mathscr{X} = \Phi(T)$ can be replicated by investing in $y_t = \partial F/\partial x\,(t, S(t))$ shares of stock and $b_t = F(t, S(t)) - y_t S(t)$ position in the risk-free asset. By definition, the value $V(t) = b_t + y_t S(t) = F(t, S(t))$ replicates the claim at time T by virtue of F's terminal condition. We thus only need to check that V satisfies the self-financing condition.

By Itô's formula for the Black-Scholes model, the Black-Scholes PDE, and the definition of b_t and y_t,

$$\begin{aligned}
dV(t) &= dF(t, S(t)) \\
&= \frac{\partial F}{\partial t}(t, S(t))\,dt + \frac{\partial F}{\partial x}(t, S(t))\,dS(t) + \frac{1}{2}\frac{\partial^2 F}{\partial x^2}(t, S(t))\,\sigma^2 S(t)^2\,du \\
&= rF(t, S(t))\,dt - rS(t)\frac{\partial F}{\partial x}(t, S(t)) - \frac{1}{2}\sigma^2 S(t)^2\frac{\partial^2 F}{\partial x^2}(t, S(t)) \\
&\quad + \frac{\partial F}{\partial x}(t, S(t))\,dS(t) + \frac{1}{2}\frac{\partial^2 F}{\partial x^2}(t, S(t))\,\sigma^2 S(t)^2\,du \\
&= r(b_t + y_t S(t))\,dt - r y_t S(t)\,dt + y_t dS(t) \\
&= r b_t dt + y_t dS(t).
\end{aligned}$$

This is the self-financing condition for the hedging portfolio V. The main pricing and hedging theorem is thus justified.

Sensitivity Analysis for Time-Inhomogeneous Lévy Process: A Malliavin Calculus Approach and Numerics

M'hamed Eddahbi and Sidi Mohamed Lalaoui Ben Cherif

Abstract The main goal of this paper is to study sensitivity analysis, with respect to the parameters of the model, in the framework of time-inhomogeneous Lévy process. This is a slight generalization of recent results of Fournié et al. (Finance Stochast 3(4):391–412, 1999 [9]), El-Khatib and Privault (Finance Stochast 8(2):161–179, 2004 [7]), Bally et al. (Ann Appl Probab 17(1):33–66, 2007 [1]), Davis and Johansson (Stochast Process Appl 116(1):101–129, 2006 [5]), Petrou (Electron J Probab 13(27):852–879, 2008 [12]), Benth et al. (Commun Stochast Anal 5(2):285–307, 2011 [2]) and El-Khatib and Hatemi (J Statist Appl Probab 3(1):171–182, 2012 [8]), using Malliavin calculus developed by Yablonski (Rocky Mountain J Math 38:669–701, 2008 [16]). This relatively recent result will help us to provide tools that are necessary for the calculation of the sensitivities. We provide some simple examples to illustrate the results achieved. In particular, we discussed the time-inhomogeneous versions of the Merton model and the Bates model.

Keywords Additive processes · Time-inhomogeneous lévy process · Malliavin calculus · Integration by parts formula · Sensitivity analysis

Mathematics Subject Classification 2010 60H07 · 60H35 · 60G51

M. Eddahbi (✉) · S.M. Lalaoui Ben Cherif
Faculty of Sciences and Techniques, Department of Mathematics, Cadi Ayyad University, B.P. 549 Marrakech, Morocco
e-mail: m.eddahbi@uca.ma

S.M. Lalaoui Ben Cherif
Faculty of Sciences Semlalia, Department of Mathematics, Cadi Ayyad University, B.P. 2390 Marrakesh, Morocco
e-mail: mohamed.lalaoui@ced.uca.ma

M. Eddahbi et al. (eds.), *Statistical Methods and Applications in Insurance and Finance*, Springer Proceedings in Mathematics & Statistics 158,
DOI 10.1007/978-3-319-30417-5_2

1 Introduction

A trader selling a financial product to a customer usually tends to avoid any risk involved in that product and therefore wants to get rid of these risks by hedging. In some cases we can make use of a static hedge and we can hedge—and—forget it, additionally we can calculate the price from the products used for hedging. But for most options this is not possible and we have to use a dynamic hedging strategy. The price sensitivities with respect to the model parameters—the Greeks—are vital inputs in this context.

The Greeks are calculated as differentials of the derivative price, which can be expressed as an expectation (in risk—neutral measure) of the discounted payoff. The Greeks are traditionally estimated by means of a finite difference approximation. This approximation contains two errors: one on the approximation of the derivative function by means of its finite difference and another one on the numerical computation of the expectation. In addition the theoretical convergence rates for finite difference approximations are not met for discontinuous payoff functions.

Fournié et al. [9] propose a method with faster convergence which consists in shifting the differential operator from the payoff functional to the diffusion kernel, introducing a weighting function. The main idea is the use of the Malliavin integration by parts formula to transform the problem of calculating derivatives by finite difference approximations to calculating expectations of the form

$$\mathrm{E}[H(S_T)\pi \,|\, S_0 = x]$$

where the weight π is a random variable and the underling price process is a Markov diffusion given by:

$$dS_t = b(S_t)dt + \sigma(S_t)dW_t, \quad S_0 = x.$$

There have been several studies that attempt to produce similar results for markets governed by processes with jumps. We mention León et al. [10], have approximated a jump—diffusion model for a simple Lévy process, and hedged an european call option using a Malliavin Calculus approach. El-Khatib and Privault [7] where a market generated by Poisson processes is considered. Their setup allows for random jump sizes, and by imposing a regularity condition on the payoff they use Malliavin calculus on Poisson space to derive weights for Asian options. Bally et al. [1] reduce the problem to a setting in which only 'finite—dimensional' Malliavin calculus is required in the case where stochastic differential equations are driven by Brownian motion and compound Poisson components. Davis and Johansson [5] have developed the Malliavin calculus for simple Lévy process which allows them to calculate the Greeks in a jump diffusion setting which satisfy a separability condition. Petrou [12] has calculated the sensitivities using Malliavin Calculus for markets generated by square integrable Lévy processes which is a extension of the paper [9]. Benth et al. [2] studied the computation of the deltas in model variation within a jump—diffusion framework with two approaches, the Malliavin calculs technics and the Fourier method. El-Khatib and Hatemi [8] estimated the price sensitivities of a trading position with regard to underlying factors in jump—diffusion models using jump times Poisson noise.

While Lévy processes offer nice features in terms of analytical tractability, the constraints of independence and stationarity of their increments prove to be rather restrictive. On one hand, the stationarity of increments of Lévy processes leads to rigid scaling properties for marginal distributions of returns, which are not observed in empirical time series of returns. On the other hand, from the point of view of risk neutral modeling, the Lévy models allow to reproduce the phenomenon of volatility smile for a given maturity, but it becomes more complicated when one tries to stick to several maturities. The inhomogeneity in time increments can improve it, hence the importance of introducing the additive processes in financial modeling. Each of the previous papers has its advantages in specific cases. However, they can only treat subclasses of Lévy processes except that of [12] but in time-homogeneous case setting.

The objective of this work is to derive stochastic weights in order to compute the Greeks in market models with jump when the discontinuity is described by a Poisson random measure with *time-inhomogeneous* intensity and then to use different numerical methods to compare the results for simpler time dependent models. The main tool uses Malliavin calculus, developed by Yablonski [16] for additive processes, that will be presented shortly at the appendix of the present document for the sake of completeness. Essentially, we introduce the time-inhomogeneity in the jump component of the risky asset price. In particular, we focus on a class of models in which the price of the underlying asset is governed by the following stochastic differential equation:

$$\begin{cases} dS_t = b(t, S_{t-})dt + \sigma(t, S_{t-})dW_t \\ \qquad\quad + \int_{\mathbb{R}_0^d} \varphi(t, S_{t-}, z)\widetilde{N}(dt, dz), \\ S_0 = x \end{cases} \tag{1}$$

where $\mathbb{R}_0^d := \mathbb{R}^d \setminus \{0_{\mathbb{R}^d}\}$, $x = (x_i)_{1 \le i \le d} \in \mathbb{R}^d$. The functions $b : \mathbb{R}^+ \times \mathbb{R}^d \longrightarrow \mathbb{R}^d$, $\sigma : \mathbb{R}^+ \times \mathbb{R}^d \longrightarrow \mathbb{R}^{d \times d}$ and $\varphi : \mathbb{R}^+ \times \mathbb{R}^d \times \mathbb{R}^d \longrightarrow \mathbb{R}^{d \times d}$, are continuously differentiable with bounded derivatives. Here

$$W_t = (W_1(t), \dots, W_d(t))$$

is a d-dimensional standard Brownian motion and

$$\widetilde{N}(dt, dz)^\top = (N_1(dt, dz_1) - \nu_t^1(z_1), \dots, N_d(dt, dz_d) - \nu_t^d(z_d))$$

where $N_k, k = 1, \dots, d$ are independent Poisson random measures on $[0, T] \times \mathbb{R}_0$, $\mathbb{R}_0 := \mathbb{R}_0^1$, with *time-inhomogeneous* Lévy measures $\nu_t^k, k = 1, \dots, d$ coming from d independent one-dimensional *time-inhomogeneous* Lévy processes. The family of positive measures $(\nu_t^k)_{1 \le k \le d}$ satisfies

$$\sum_{k=1}^{d}\int_0^T\int_{\mathbb{R}_0}(|z_k|^2\wedge 1)\nu_t^k(dz_k)dt<\infty$$

and $\nu_t^k(\{0\})=0, k=1,\dots,d$. Let $b(t,x)=b_i(t,x))_{1\le i\le d}$, $\sigma(t,x)=\sigma_{ij}(t,x)_{1\le i\le d,1\le j\le d}$ and $\varphi(t,x,z)=\varphi_{ik}(t,x,z)_{1\le i\le d,1\le k\le d}$ be the coefficients of (1) in the component form. Then $S_t=(S_i(t))_{1\le i\le d}$ in (1) can be equivalently written as

$$\begin{cases} dS_t^i=b_i(t,S_{t-})dt+\sum_{j=1}^d\sigma_{ij}(t,S_{t-})dW_j(t)\\ \qquad +\sum_{k=1}^d\int_{\mathbb{R}_0}\varphi_{ik}(t,S_{t-},z_k)\widetilde{N}_k(dt,dz_k),\\ S_0^i=x_i. \end{cases}\tag{2}$$

To guarantee a unique strong solution to (1), we assume that the coefficients of (1) satisfy linear growth and Lipschitz continuity, i.e.,

$$\|b(t,x)\|^2+\|\sigma(t,x)\|^2+\sum_{k=1}^d\sum_{i=1}^d\int_{\mathbb{R}_0}|\varphi_{ik}(t,x,z_k)|^2\nu_t^k(dz_k)\le C(1+\|x\|^2)\tag{3}$$

and

$$\|b(t,x)-b(t,y)\|^2+\|\sigma(t,x)-\sigma(t,y)\|^2\le K_1\|x-y\|^2\tag{4}$$

for all $x,y\in\mathbb{R}^d$ and $t\in[0,T]$, with C and K_1 are positive constants.
We suppose that there exists a family of functions $\rho_k:\mathbb{R}\longrightarrow\mathbb{R}, k=1,\dots,d$ such that

$$\sup_{0\le t\le T}\int_{\mathbb{R}_0}\sum_{k=1}^d|\rho_k(z_k)|^2\nu_t^k(dz_k)<\infty,\tag{5}$$

and a positive constant K_2 such that

$$\sum_{i=1}^d|\varphi_{ik}(t,x,z_k)-\varphi_{ik}(t,y,z_k)|^2\le K_2|\rho(z_k)|^2\|x-y\|^2,\tag{6}$$

for all $x,y\in\mathbb{R}^d, t\in[0,T]$ and $z_k\in\mathbb{R}, k=1,\dots,d$. Similarly to the homogeneous case, we have the following lemma:

Lemma 1.1 *Under the above conditions there exists a unique solution $(S_t)_{t\in[0,T]}$ for (1). Moreover, there exists a positive constant C_0 such that*

$$\mathrm{E}\left[\sup_{0\le t\le T}\|S_t\|^2\right]<C_0.$$

2 Regularity of Solutions of SDEs Driven by Time-Inhomogeneous Lévy Processes

The aim of this section is to prove that under specific conditions the solution of a stochastic differential equation belongs to the domains $\mathbb{D}^{1,2}$ (see Sects. 4.14 and 4.16). Having in mind the applications in finance, we will also provide a specific expression for the Wiener directional derivative of the solution.

Remark 2.1 The theory developed in the Appendix also holds in the case that our space is generated by an d-dimensional Wiener process and d-dimensional random Poisson measures. However, we will have to introduce new notation for the directional derivatives in order to simplify things. For the multidimensional case,

$$D_{t,0} = (D^{(1)}_{t,0}, \dots, D^{(d)}_{t,0})$$

will denote a row vector, where the element $D^{(j)}_{t,0}$ of the jth row is the directional derivative for the Wiener process W_j, for all $j = 1, \dots, d$. Similarly, for all $z = (z_k)_{1 \le k \le d} \in \mathbb{R}^d_0$ we define the row vector

$$D_{t,z} = (D^{(1)}_{t,z_1}, \dots, D^{(d)}_{t,z_d})$$

where the element $D^{(k)}_{t,z_k}$ of the kth row is the directional derivative for the random Poisson measure $\widetilde{N}_k$, for all $k = 1, \dots, d$. For what follows we denote with σ_j the jth column vector of σ and φ_k the kth column vector of φ.

Theorem 2.2 *Let $(S_t)_{t\in[0,T]}$ be the solution of (1). Then $S_t \in \mathbb{D}^{1,2}$ for all $t \in [0, T]$, and we have*

1. *The Malliavin derivative $D^{(j)}_{r,0} S_t$ with respect to W_j satisfies the following linear equation:*

$$\begin{aligned} D^{(j)}_{r,0} S_t &= \sum_{i=1}^{d} \int_r^t \frac{\partial b}{\partial x_i}(u, S_{u-}) D^{(j)}_{r,0} S^i_{u-} du + \sigma_j(r, S_{r-}) \\ &+ \sum_{i=1}^{d} \sum_{\alpha=1}^{d} \int_r^t \frac{\partial \sigma_\alpha}{\partial x_i}(u, S_{u-}) D^{(j)}_{r,0} S^i_{u-} dW_\alpha(u) \\ &+ \sum_{i=1}^{d} \int_r^t \int_{\mathbb{R}^d_0} \frac{\partial \varphi}{\partial x_i}(u, S_{u-}, y) D^{(j)}_{r,0} S^i_{u-} \widetilde{N}(du, dy), \end{aligned}$$

for $0 \le r \le t$ a.e. and $D^{(j)}_{r,0} S_t = 0$ a.e. otherwise.

2. *For all $z \in \mathbb{R}^d_0$, The Malliavin derivative $D_{r,z} S_t$ with respect to $\widetilde{N}$ satisfies the following linear equation:*

$$D_{r,z}S_t = \int_r^t D_{r,z}b(u, S_{u-})du + \int_r^t D_{r,z}\sigma(u, S_{u-})dW_u$$
$$+\varphi(r, S_{r-}, z) + \int_r^t \int_{\mathbb{R}_0^d} D_{r,z}\varphi(u, S_{u-}, y)\widetilde{N}(du, dy),$$

for $0 \le r \le t$ *a.e. and* $D_{r,z}S_t = 0$ *a.e. otherwise.*

Proof 1. We consider the Picard approximations S_t^n, $n \geq 0$, given by

$$\begin{cases} S_t^0 = x \\ S_t^{n+1} = x + \int_0^t b(u, S_{u-}^n)du + \int_0^t \sigma(u, S_{u-}^n)dW_u \\ \quad + \int_0^t \int_{\mathbb{R}_0^d} \varphi(u, S_{u-}^n, z)\widetilde{N}(du, dz). \end{cases} \tag{7}$$

From Lemma 1.1 we know that

$$\mathrm{E}\left[\sup_{0 \le t \le T} |S_t^n - S_t|^2\right] \underset{n\to\infty}{\longrightarrow} 0.$$

By induction, we prove that the following statements hold true for all $n \geq 0$.
Hypothesis (H)

(a) $S_t^n \in \mathbb{D}^{1,2}$ for all $t \in [0, T]$.

(b) $\xi_n(t) = \sup_{0 \le r \le t} \mathrm{E}\left[\sup_{r \le u \le t} \left|D_{r,0}S_u^n\right|^2\right] < \infty.$

(c) $\xi_{n+1}(t) \le \alpha + \beta \int_0^t \xi_n(u)du$ for some constants α, β.

For $n = 0$, it is straightforward that **(H)** is satisfied. Assume that **(H)** holds for a certain n. We would prove it for $n + 1$. By Proposition 4.12 $b(u, S_{u-}^n)$, $\sigma(u, S_{u-}^n)$ and $\varphi(u, S_{u-}^n, z)$ are in $\mathbb{D}^{1,2}$. Furthermore,

$$D_{r,0}b_i(u, S_{u-}^n) = \sum_{\alpha=1}^d \frac{\partial b_i}{\partial x_\alpha}(u, S_{u-}^n)D_{r,0}S_{u-}^{n,\alpha}\mathbf{1}_{\{r\le u\}},$$

$$D_{r,0}\sigma_{ij}(u, S_{u-}^n) = \sum_{\alpha=1}^d \frac{\partial \sigma_{ij}}{\partial x_\alpha}(u, S_{u-}^n)D_{r,0}S_{u-}^{n,\alpha}\mathbf{1}_{\{r\le u\}},$$

$$D_{r,0}\varphi_{ik}(u, S_{u-}^n, z_k) = \sum_{\alpha=1}^d \frac{\partial \varphi_{ik}}{\partial x_\alpha}(u, S_{u-}^n, z_k)D_{r,0}S_{u-}^{n,\alpha}\mathbf{1}_{\{r\le u\}}.$$

Since the functions b, σ and φ are continuously differentiable with bounded first derivatives in the second direction and taking into account the conditions (4) and (6) we have

$$\begin{aligned}\left\|D_{r,0}b_i(u,S^n_{u-})\right\|^2 &\le K_1\left\|D_{r,0}S^n_{u-}\right\|^2,\\ \left\|D_{r,0}\sigma_{ij}(u,S^n_{u-})\right\|^2 &\le K_1\left\|D_{r,0}S^n_{u-}\right\|^2,\\ \left\|D_{r,0}\varphi_{ik}(u,S^n_{u-},z_k)\right\|^2 &\le K_2|\rho(z_k)|^2\left\|D_{r,0}S^n_{u-}\right\|^2.\end{aligned} \tag{8}$$

However, $\int_0^t b(u,S^n_{u-})du$, $\int_0^t \sigma(u,S^n_{u-})dW_u$ and $\int_0^t\int_{\mathbb{R}_0^d}\varphi(u,S^n_{u-},z)\widetilde{N}(dt,dz)$ are in $\mathbb{D}^{1,2}$. Which implies that S^{n+1}_t to $\mathbb{D}^{1,2}$ and we have

$$\begin{aligned}D^{(j)}_{r,0}\int_0^t b_i(u,S^n_{u-})du &= \int_r^t D^{(j)}_{r,0}b_i(u,S^n_{u-})du,\\ D^{(j)}_{r,0}\sum_{\alpha=1}^d\int_0^t\sigma_{i\alpha}(u,S^n_{u-})dW^\alpha_u &= \sigma_{ij}(r,S^n_{r-})+\sum_{\alpha=1}^d\int_r^t D^{(j)}_{r,0}\sigma_{i\alpha}(u,S^n_{u-})dW_\alpha(u),\\ D^{(j)}_{r,0}\sum_{k=1}^d\int_0^t\int_{\mathbb{R}_0}\varphi_{ik}(u,S^n_{u-},z_k)\widetilde{N}_k(dt,dz_k) &= \sum_{k=1}^d\int_r^t\int_{\mathbb{R}_0}D^{(j)}_{r,0}\varphi_{ik}(u,S^n_{u-},z_k)\widetilde{N}_k(dt,dz_k).\end{aligned}$$

Thus

$$\begin{aligned}D^{(j)}_{r,0}S^{n+1}_t &= \int_r^t D^{(j)}_{r,0}b(u,S^n_{u-})du+\sigma_j(r,S^n_{r-})+\sum_{\alpha=1}^d\int_r^t D^{(j)}_{r,0}\sigma_\alpha(u,S^n_{u-})dW_\alpha(u)\\ &\quad+\sum_{k=1}^d\int_r^t\int_{\mathbb{R}_0}D^{(j)}_{r,0}\varphi_k(u,S^n_{u-},z_k)\widetilde{N}_k(dt,dz_k).\end{aligned}$$

We conclude that

$$\begin{aligned}\mathrm{E}\left[\sup_{r\le v\le t}|D^{(j)}_{r,0}S^{n+1}_v|^2\right] &\le 4\left\{\mathrm{E}\left[\sup_{r\le v\le t}\left|\int_r^v D^{(j)}_{r,0}b(u,S^n_{u-})du\right|^2\right]\right.\\ &+\mathrm{E}\left[\sup_{0\le t\le T}|\sigma_j(t,S^n_t)|^2\right]+\mathrm{E}\left[\sup_{r\le v\le t}\left|\sum_{\alpha=1}^d\int_r^v D^{(j)}_{r,0}\sigma_\alpha(u,S^n_{u-})dW_\alpha(u)\right|^2\right]\\ &\left.+\mathrm{E}\left[\sup_{r\le v\le t}\left|\sum_{k=1}^d\int_r^v\int_{\mathbb{R}_0}D^{(j)}_{r,0}\varphi_k(u,S^n_{u-},z_k)\widetilde{N}_k(dt,dz_k)\right|^2\right]\right\}.\end{aligned}$$

Using Cauchy–Schwarz inequality and Burkholder–Davis–Gundy inequality (see [14], Theorem 48 p. 193), there exists a constant $K>0$ such that

$$\begin{aligned}\mathrm{E}\left[\sup_{r\le v\le t}|D^{(j)}_{r,0}S^{n+1}_v|^2\right] &\le K\left\{(t-r)\mathrm{E}\left[\int_r^t|D^{(j)}_{r,0}b(u,S^n_{u-})|^2du\right]\right.\\ &+\mathrm{E}\left[\sup_{0\le t\le T}|\sigma_j(t,S^n_t)|^2\right]+\mathrm{E}\left[\sum_{\alpha=1}^d\int_r^t|D^{(j)}_{r,0}\sigma_\alpha(u,S^n_{u-})|^2du\right]\end{aligned}$$

$$+ \mathrm{E}\left[\sum_{k=1}^{d}\int_{r}^{v}\int_{\mathbb{R}_0}|D_{r,0}^{(j)}\varphi_k(u,S_{u-}^n,z_k)|^2\nu_u^k(dz_k)du\right]\Bigg\}.$$

From (6) and (8) we reach

$$\mathrm{E}\left[\sup_{r\le u\le t}|D_{r,0}^{(j)}S_u^{n+1}|^2\right]\le K\mathrm{E}\left[\sup_{0\le t\le T}|\sigma_j(t,S_t^n)|^2\right]$$
$$+K\left(K_1(T+1)+K_2\sup_{0\le t\le T}\int_{\mathbb{R}_0}\sum_{k=1}^{d}|\rho_k(z_k)|^2\nu_t^k(dz_k)\right)\int_r^t\mathrm{E}\left[|D_{r,0}^{(j)}S_{u-}^n|^2\right]du.$$

Then, from (3)

$$\mathrm{E}\left[\sup_{r\le u\le t}|D_{r,0}^{(j)}S_u^{n+1}|^2\right]\le KC\left(1+\mathrm{E}\left[\sup_{0\le t\le T}|S_t^n|^2\right]\right)$$
$$+K\left(K_1(T+1)+K_2\sup_{0\le t\le T}\int_{\mathbb{R}_0}\sum_{k=1}^{d}|\rho_k(z_k)|^2\nu_t^k(dz_k)\right)\int_r^t\mathrm{E}\left[\sup_{r\le v\le u}|D_{r,0}^{(j)}S_{v-}^n|^2\right]du.$$

Consequently

$$\xi_{n+1}(t)\le\alpha+\beta\int_0^t\xi_n(u)du,$$

where

$$\alpha:=KC\left(1+\sup_{n\in\mathbb{N}}\mathrm{E}\left[\sup_{0\le t\le T}|S_t^n|^2\right]\right)<\infty$$

and, using (5), we have

$$\beta:=K\left(K_1(T+1)+K_2+\sup_{0\le t\le T}\int_{\mathbb{R}_0}\sum_{k=1}^{d}|\rho_k(z_k)|^2\nu_t^k(dz_k)\right)<\infty.$$

By induction, we can easily prove that, for all $n\in\mathbb{N}$ and $t\in[0,T]$

$$\xi_n(t)\le\alpha\sum_{i=0}^{n}\frac{(\beta t)^i}{i!}.$$

Hence, for all $n\in\mathbb{N}$ and $t\in[0,T]$

$$\xi_n(t)\le\alpha e^{\beta t}<\infty,$$

which implies that the derivatives of S_t^n are bounded in $L^2(\Omega \times [0,T])$ uniformly in n. Hence, we deduce that the random variable S_t belongs to $\mathbb{D}^{1,2}$ and by applying the chain rule to (1) we achieve our proof.

2. Following the same steps we can prove the second claim of the theorem.

As in the classical Malliavin calculus we are able to associate the solution of (1) with the first variation process $Y_t := \nabla_x S_t$. We reach the following proposition which provides us with a simpler expression for $D_{r,0}S_t$.

Proposition 2.3 *Let $(S_t)_{t\in[0,T]}$ be the solution of (1). Then the derivative satisfies the following equation:*

$$D_{r,0}S_t = Y_t Y_{r-}^{-1}\sigma(r, S_{r-})\mathbf{1}_{\{r\leq t\}} \quad a.e. \tag{9}$$

where $(Y_t)_t$ is the first variation process of $(S_t)_t$.

Proof Let $(S_t)_{t\in[0,T]}$ be the solution of (1). Then

$$\begin{aligned}
D_{r,0}^{(j)}S_t^i &= \sum_{\beta=1}^d \int_r^t \frac{\partial b_i}{\partial x_\beta}(u, S_{u-})D_{r,0}^{(j)}S_{u-}^\beta du + \sigma_{ij}(r, S_{r-}) \\
&+ \sum_{\beta=1}^d\sum_{\alpha=1}^d \int_r^t \frac{\partial \sigma_{i\alpha}}{\partial x_\beta}(u, S_{u-})D_{r,0}^{(j)}S_{u-}^\beta dW_\alpha(u) \\
&+ \sum_{\beta=1}^d\sum_{k=1}^d \int_r^t\int_{\mathbb{R}_0} \frac{\partial \varphi_{ik}}{\partial x_\beta}(u, S_{u-}, z_k)D_{r,0}^{(j)}S_{u-}^\beta \widetilde{N}_k(du, dz_k).
\end{aligned}$$

The $d \times d$ matrix–valued process Y_t is given by

$$\begin{aligned}
Y_t^{ij} &:= \frac{\partial S_t^i}{\partial x_j} \\
&= \delta_{ij} + \sum_{k=1}^d \int_0^t \frac{\partial b_i}{\partial x_k}(u, S_{u-})Y_{u-}^{kj} du \\
&+ \sum_{k=1}^d\sum_{\alpha=1}^d \int_0^t \frac{\partial \sigma_{i\alpha}}{\partial x_k}(u, S_{u-})Y_{u-}^{kj} dW_\alpha(u) \\
&+ \sum_{k=1}^d\sum_{\beta=1}^d \int_0^t\int_{\mathbb{R}_0} \frac{\partial \varphi_{i\beta}}{\partial x_k}(u, S_{u-}, z_\beta)Y_{u-}^{kj} \widetilde{N}_\beta(du, dz_\beta)
\end{aligned}$$

with $\delta_{ii} = 1$ and $\delta_{ij} = 0$ if $i \neq j$. Let $(Z_t)_{0\leq t\leq T}$ be a $d \times d$ matrix–valued process that satisfies the following equation

$$
\begin{aligned}
Z_t^{ij} &= \delta_{ij} + \sum_{k=1}^{d} \int_0^t \left(-\frac{\partial b_k}{\partial x_j}(u, S_{u-}) + \sum_{n=1}^{d} \sum_{\alpha=1}^{d} \frac{\partial \sigma_{k\alpha}}{\partial x_n}(u, S_{u-}) \frac{\partial \sigma_{n\alpha}}{\partial x_j}(u, S_{u-}) \right) Z_{u-}^{ik} du \\
&+ \sum_{k=1}^{d} \sum_{\beta=1}^{d} \int_0^t \int_{\mathbb{R}_0} \frac{\sum_{n=1}^{d} \frac{\partial \varphi_{k\beta}}{\partial x_n}(u, S_{u-}, z_\beta) \frac{\partial \varphi_{n\beta}}{\partial x_j}(u, S_{u-}, z_\beta)}{1 + \frac{\partial \varphi_{k\beta}}{\partial x_j}(u, S_{u-}, z_\beta)} Z_{u-}^{ik} \nu_u^\beta (dz_\beta) du \\
&- \sum_{k=1}^{d} \sum_{\alpha=1}^{d} \int_0^t \frac{\partial \sigma_{k\alpha}}{\partial x_j}(u, S_{u-}) Z_{u-}^{ik} dW_\alpha(u) \\
&- \sum_{k=1}^{d} \sum_{\beta=1}^{d} \int_0^t \int_{\mathbb{R}_0} \frac{\frac{\partial \varphi_{k\beta}}{\partial x_j}(u, S_{u-}, z_\beta)}{1 + \frac{\partial \varphi_{k\beta}}{\partial x_j}(u, S_{u-}, z_\beta)} Z_{u-}^{ik} \widetilde{N}_\beta(du, dz_\beta).
\end{aligned}
$$

By means of Itô's formula, one can check that

$$
\sum_{j=1}^{d} Z_t^{ij} Y_t^{jk} = \delta_{ik}.
$$

Hence $Z_t Y_t = Z_t Y_t = I_d$ where I_d is the unit matrix of size d. As a consequence, for any $t \geq 0$ the matrix Y_t is invertible and $Y_t^{-1} = Z_t$. Applying again Itô's formula, it holds that

$$
D_{r,0}^{(j)} S_t^i = \sum_{n=1}^{d} \sum_{k=1}^{d} Y_t^{ik} Z_r^{kn} \sigma_{nj}(r, S_{r-}) \quad \text{for all} \quad r \leq t.
$$

Then the result follows.

2.1 Greeks

For $n \in \mathbb{N}^*$ we define the payoff $H := H(S_{t_1}, S_{t_2}, \ldots, S_{t_n})$ to be a square integrable function discounted from maturity T and evaluated at the times $t_1, t_2, \ldots, t_n$ with the convention that $t_0 = 0$ and $t_n = T$. Under a chosen, since we do not have uniqueness, risk neutral measure, denoted by $\mathbb{Q}$, the price $\mathscr{C}(x)$ of the contingent claim given an initial value is then expressed as:

$$
\mathscr{C}(x) = \mathrm{E}_{\mathbb{Q}} \left[H(S_{t_1}, S_{t_2}, \ldots, S_{t_n}) \right].
$$

In what follows, we assume the next ellipticity[1] condition for the diffusion matrix σ.

---

[1]This is to ensure that we can find some solutions for the weighting functions, since it often requires to take the inverse of the volatility function.

Assumption 2.4 The diffusion matrix σ satisfies the uniform ellipticity condition:

$$\exists\, \eta > 0 \;\; \xi^* \sigma^*(t,x)\sigma(t,x)\xi > \eta \|\xi\|^2, \quad \forall\, \xi, x \in \mathbb{R}^d.$$

Using the Malliavin calculus developed in the Sect. 2.1 we are able to calculate the Greeks for the one–dimensional process $(S_t)_{t\in[0,T]}$ that satisfies equation (1).

2.2 Variation in the Initial Condition

In this section, we provide an expression for the derivatives of the expectation $\mathscr{C}(x)$ with respect to the initial condition x in the form of a weighted expectation of the same functional.

Let us define the set:

$$T_n = \left\{ a \in L^2([0,T]) : \int_0^{t_i} a(u)du = 1 \;\; \forall\, i = 1, 2, \ldots, n \right\}$$

where t_i, $i = 1, 2, \ldots, n$ are as defined in the Sect. 2.1.

Proposition 2.5 *Assume that the diffusion matrix σ is uniformly elliptic. Then for all $a \in T_n$,*

$$\nabla_x \mathscr{C}(x) = \mathrm{E}_{\mathbb{Q}} \left[H(S_{t_1}, S_{t_2}, \ldots, S_{t_n}) \int_0^T a(u)\sigma^{-1}(u, S_{u-})Y_{u-} dW_u \right].$$

Proof Let H be a continuously differentiable function with bounded gradient. Then we can differentiate inside the expectation (see Fournié et al. [9] for details) and we have

$$\begin{aligned}
\nabla_x \mathscr{C}(x) &= \mathrm{E}_{\mathbb{Q}} \left[\sum_{i=1}^{n} \nabla_i H(S_{t_1}, S_{t_2}, \ldots, S_{t_n}) \nabla_x S_{t_i} \right] \\
&= \mathrm{E}_{\mathbb{Q}} \left[\sum_{i=1}^{n} \nabla_i H(S_{t_1}, S_{t_2}, \ldots, S_{t_n}) Y_{t_i} \right]
\end{aligned}$$

where $\nabla_i H(S_{t_1}, S_{t_2}, \ldots, S_{t_n})$ is the gradient of H with respect to S_{t_i} for $i = 1, \ldots, n$. For any $a \in T_n$ and $i = 1, \ldots, n$ and using (9) we find

$$Y_{t_i} = \int_0^T a(u) D_{u,0} S_{t_i} \sigma^{-1}(u, S_{u-}) Y_{u-} du.$$

From Proposition 4.12 we reach

$$\begin{aligned}
\nabla_x \mathscr{C}(x) &= \mathrm{E}_{\mathbb{Q}}\left[\int_0^T \sum_{i=1}^n \nabla_i H(S_{t_1}, S_{t_2}, \ldots, S_{t_n}) a(u) D_{u,0} S_{t_i} \sigma^{-1}(u, S_{u-}) Y_{u-} du\right] \\
&= \mathrm{E}_{\mathbb{Q}}\left[\int_0^T D_{u,0} H(S_{t_1}, S_{t_2}, \ldots, S_{t_n}) a(u) \sigma^{-1}(u, S_{u-}) Y_{u-} du\right] \\
&= \mathrm{E}_{\mathbb{Q}}\left[\int_0^T \int_{\mathbb{R}} D_{u,z} H(S_{t_1}, S_{t_2}, \ldots, S_{t_n}) a(u) \sigma^{-1}(u, S_{u-}) Y_{u-} du \delta_0(dz)\right].
\end{aligned}$$

Into measure $\pi(dudz)$ defined in Sect. 4.12 we replace Δ by 0 and $\mu(du)$ by a Lebesgue measure du. Then

$$\nabla_x \mathscr{C}(x) = \mathrm{E}_{\mathbb{Q}}\left[\int_0^T \int_{\mathbb{R}} D_{u,z} H(S_{t_1}, S_{t_2}, \ldots, S_{t_n}) a(u) \sigma^{-1}(u, S_{u-}) Y_{u-} \mathbf{1}_{\{0\}}(z) \pi(dudz)\right].$$

Using the integration by parts formula (see Sect. 4.14), we have

$$\nabla_x \mathscr{C}(x) = \mathrm{E}_{\mathbb{Q}}\left[H(S_{t_1}, S_{t_2}, \ldots, S_{t_n}) \delta\left(a(\cdot)\sigma^{-1}(\cdot, S_{\cdot}) Y_{\cdot} \mathbf{1}_{\{z=0\}}(\cdot)\right)\right].$$

However, $\left(a(u)\sigma^{-1}(t, S_{t-}) Y_{t-}\right)_{0 \le t \le T}$ is a predictable process, thus the Skorohod integral coincides with the Itô stochastic integral.

$$\nabla_x \mathscr{C}(x) = \mathrm{E}_{\mathbb{Q}}\left[H(S_{t_1}, S_{t_2}, \ldots, S_{t_n}) \int_0^T a(u) \sigma^{-1}(u, S_{u-}) Y_{u-} dW_u\right].$$

Since the family of continuously differentiable functions is dense in L^2, the result hold for any $H \in L^2$ (see Fournié et al. [9] for details).

2.3 Variation in the Drift Coefficient

Let $\widetilde{b} : \mathbb{R}^+ \times \mathbb{R}^d \longrightarrow \mathbb{R}^d$ be a function such that for every $\varepsilon \in [-1, 1]$, $\widetilde{b}$ and $b + \varepsilon \widetilde{b}$ are continuously differentiable with bounded first derivatives in the space directions.

We then define the drift–perturbed process $(S_t^\varepsilon)_t$ as a solution of the following perturbed stochastic differential equation:

$$\begin{cases} dS_t^\varepsilon = (b(t, S_{t-}^\varepsilon) + \varepsilon \widetilde{b}(t, S_{t-}^\varepsilon))dt + \sigma(t, S_{t-}^\varepsilon) dW_t \\ \qquad + \int_{\mathbb{R}_0^d} \varphi(t, S_{t-}^\varepsilon, z) \widetilde{N}(dt, dz), \text{ with } S_0^\varepsilon = x. \end{cases} \tag{10}$$

We can relate to this perturbed process the perturbed price $\mathscr{C}^\varepsilon(x)$ defined by

$$\mathscr{C}^\varepsilon(x) = \mathrm{E}_{\mathbb{Q}}\left[H(S_{t_1}^\varepsilon, S_{t_1}^\varepsilon, \ldots, S_{t_n}^\varepsilon)\right].$$

Proposition 2.6 *Assume that the diffusion matrix σ is uniformly elliptic. Then we have*

$$Rho = \frac{\partial \mathscr{C}^\varepsilon}{\partial \varepsilon}(x)\bigg|_{\varepsilon=0} = \mathrm{E}_{\mathbb{Q}}\left[H(S_{t_1}, S_{t_2}, \ldots, S_{t_n}) \int_0^T (\sigma^{-1}\widetilde{b})(t, S_{t-})dW_t\right].$$

Proof We introduce the random variable

$$\widetilde{D}_T^\varepsilon = \exp\left(\varepsilon \int_0^T (\sigma^{-1}\widetilde{b})(t, S_{t-}^\varepsilon)dW_t - \frac{\varepsilon^2}{2}\int_0^T \|(\sigma^{-1}\widetilde{b})(t, S_{t-}^\varepsilon)\|^2 dt\right).$$

The Novikov condition is satisfied since

$$\mathrm{E}_{\mathbb{Q}}\left[\exp\left(\frac{\varepsilon^2}{2}\int_0^T \|(\sigma^{-1}\widetilde{b})(t, S_{t-}^\varepsilon)\|^2 dt\right)\right] < +\infty.$$

As well as $\mathrm{E}_{\mathbb{Q}}[\widetilde{D}_T^\varepsilon] = 1$, then we can define new probability measure $\mathbb{Q}^\varepsilon$ by its Radon–Nikodym derivative with respect to the risk–neutral probability measure $\mathbb{Q}$:

$$\widetilde{D}_T^\varepsilon = \frac{d\mathbb{Q}^\varepsilon}{d\mathbb{Q}}\Big/ \mathscr{F}_T.$$

By changing of measure, we can write

$$\begin{aligned}\mathrm{E}_{\mathbb{Q}}\left[H(S_{t_1}^\varepsilon, S_{t_1}^\varepsilon, \ldots, S_{t_n}^\varepsilon)\right] &= \mathrm{E}_{\mathbb{Q}^\varepsilon}\left[H(S_{t_1}^\varepsilon, S_{t_1}^\varepsilon, \ldots, S_{t_n}^\varepsilon)\frac{d\mathbb{Q}}{d\mathbb{Q}^\varepsilon}\right] \\ &= \mathrm{E}_{\mathbb{Q}}\left[H(S_{t_1}, S_{t_2}, \ldots, S_{t_n})D_T^\varepsilon\right]\end{aligned}$$

where

$$\begin{aligned}D_T^\varepsilon &= \exp\left(-\varepsilon\int_0^T \left((\sigma^{-1}\widetilde{b})(t, S_{t-})\right)dW_t - \frac{\varepsilon^2}{2}\int_0^T \|(\sigma^{-1}\widetilde{b})(t, S_{t-})\|^2 dt\right) \\ &= 1 - \varepsilon\int_0^T \left((\sigma^{-1}\widetilde{b})(t, S_{t-})\right) D_t^\varepsilon dW_t\end{aligned}$$

which implies that

$$\Bigg|\frac{\mathrm{E}_{\mathbb{Q}}\left[H(S_{t_1}^\varepsilon, S_{t_1}^\varepsilon, \ldots, S_{t_n}^\varepsilon)\right] - \mathrm{E}_{\mathbb{Q}}\left[H(S_{t_1}, S_{t_2}, \ldots, S_{t_n})\right]}{\varepsilon} \\ - \mathrm{E}_{\mathbb{Q}}\left[H(S_{t_1}, S_{t_2}, \ldots, S_{t_n})\int_0^T (\sigma^{-1}\widetilde{b})(t, S_{t-})dW_t\right]\Bigg|^2$$

$$= \left| \mathrm{E}_{\mathbb{Q}} \left[H(S_{t_1}, S_{t_2}, \ldots, S_{t_n}) \left(\frac{D_T^\varepsilon - 1}{\varepsilon} - \left(\int_0^T (\sigma^{-1}\widetilde{b})(t, S_{t-}) dW_t \right) \right) \right] \right|^2$$
$$\leq \mathrm{E}_{\mathbb{Q}} \left[|H(S_{t_1}, S_{t_2}, \ldots, S_{t_n})|^2 \right] \mathrm{E}_{\mathbb{Q}} \left[\left| \frac{D_T^\varepsilon - 1}{\varepsilon} - \int_0^T ((\sigma^{-1}\widetilde{b})(t, S_{t-}) dW_t) \right|^2 \right].$$

2.4 Variation in the Diffusion Coefficient

In this section, we provide an expression for the derivatives of the price $\mathscr{C}(x)$ with respect to the diffusion coefficient σ. We introduce the set of deterministic functions

$$\widetilde{T}_n = \left\{ a \in L^2([0, T]) : \int_{t_{i-1}}^{t_i} a(u) du = 1 \ \ \forall\, i = 1, 2, \ldots, n \right\}$$

where $t_i, i = 1, 2, \ldots, n$ are as defined in the Sect. 2.1. Let $\widetilde{\sigma} : \mathbb{R}^+ \times \mathbb{R}^d \longrightarrow \mathbb{R}^d \times \mathbb{R}^d$ a direction function for the diffusion such that for every $\varepsilon \in [-1, 1]$, $\widetilde{\sigma}$ and $\sigma + \varepsilon\widetilde{\sigma}$ are continuously differentiable with bounded first derivatives in the second direction and verify Lipschitz conditions such that the following assumption is satisfied:

Assumption 2.7 The diffusion matrix $\sigma + \varepsilon\widetilde{\sigma}$ satisfies the uniform ellipticity condition for every $\varepsilon \in [-1, 1]$:

$$\exists\, \eta > 0 \ \ \xi^* \left(\sigma + \varepsilon\widetilde{\sigma}\right)^* (t, x) \left(\sigma + \varepsilon\widetilde{\sigma}\right) (t, x) \xi > \eta \|\xi\|^2, \quad \forall\, \xi, x \in \mathbb{R}^d.$$

We then define the diffusion–perturbed process $(S_t^{\varepsilon,\widetilde{\sigma}})_{0 \leq t \leq T}$ as a solution of the following perturbed stochastic differential equation:

$$\begin{cases} dS_t^{\varepsilon,\widetilde{\sigma}} = b(t, S_{t-}^{\varepsilon,\widetilde{\sigma}}) dt + \left(\sigma(t, S_{t-}^{\varepsilon,\widetilde{\sigma}}) + \varepsilon\widetilde{\sigma}(t, S_{t-}^{\varepsilon,\widetilde{\sigma}}) \right) dW_t \\ \qquad\qquad + \int_{\mathbb{R}_0^d} \varphi(t, S_{t-}^{\varepsilon,\widetilde{\sigma}}, z) \widetilde{N}(dt, dz), \text{ with } S_0^{\varepsilon,\widetilde{\sigma}} = x. \end{cases}$$

We can also relate to this perturbed process the perturbed price $\mathscr{C}^{\varepsilon,\widetilde{\sigma}}(x)$ defined by

$$\mathscr{C}^{\varepsilon,\widetilde{\sigma}}(x) := \mathrm{E}_{\mathbb{Q}} \left[H(S_{t_1}^{\varepsilon,\widetilde{\sigma}}, S_{t_1}^{\varepsilon,\widetilde{\sigma}}, \ldots, S_{t_n}^{\varepsilon,\widetilde{\sigma}}) \right].$$

We will need to introduce the variation process with respect to the parameter ε

$$dZ_t^{\varepsilon,\widetilde{\sigma}} = b'(t, S_{t-}^{\varepsilon,\widetilde{\sigma}}) Z_{t-}^{\varepsilon,\widetilde{\sigma}} dt + \left(\sigma'(t, S_{t-}^{\varepsilon,\widetilde{\sigma}}) + \varepsilon\widetilde{\sigma}'(t, S_{t-}^{\varepsilon,\widetilde{\sigma}}) \right) Z_{t-}^{\varepsilon,\widetilde{\sigma}} dW_t$$
$$+ \widetilde{\sigma}(t, S_{t-}^{\varepsilon,\widetilde{\sigma}}) dW_t + \int_{\mathbb{R}_0^d} \varphi'(t, S_{t-}^{\varepsilon,\widetilde{\sigma}}, z) Z_{t-}^{\varepsilon,\widetilde{\sigma}} \widetilde{N}(dt, dz) \text{ and } Z_0^{\varepsilon,\widetilde{\sigma}} = 0,$$

so that $\frac{\partial S_t^{\varepsilon,\widetilde{\sigma}}}{\partial \varepsilon} = Z_t^{\varepsilon,\widetilde{\sigma}}$. We simply use the notation S_t, Y_t and $Z_t^{\widetilde{\sigma}}$ for $S_t^{0,\widetilde{\sigma}}$, $Y_t^{0,\widetilde{\sigma}}$ and $Z_t^{0,\widetilde{\sigma}}$ where the first variation process is given by $Y_t^{0,\widetilde{\sigma}} := \nabla_x S_t^{0,\widetilde{\sigma}}$. Next, consider the process $(\beta_t^{\widetilde{\sigma}})_{t\in[0,T]}$ defined by

$$\beta_t^{\widetilde{\sigma}} := Y_t^{-1} Z_t^{\widetilde{\sigma}}, \quad 0 \le t \le T \ \ a.e.$$

Proposition 2.8 *Assume that Hypothesis 2.7 holds. Set*

$$\widetilde{\beta}_t^{a,\widetilde{\sigma}} = \sum_{i=1}^{n} a(t)\left(\beta_{t_i}^{\widetilde{\sigma}} - \beta_{t_{i-1}}^{\widetilde{\sigma}}\right) \mathbf{1}_{[t_{i-1},t_i[}(t).$$

Suppose further that the process $(\sigma^{-1}(t, S_t) Y_t \widetilde{\beta}_t^{a,\widetilde{\sigma}} \delta_0(z))_{(t,z)}$ belongs to $Dom(\delta)$, then we have for any $a \in \widetilde{T}_n$:

$$Vega = \frac{\partial \mathscr{C}^{\varepsilon,\widetilde{\sigma}}}{\partial \varepsilon}(x)\bigg|_{\varepsilon=0} = \mathrm{E}_{\mathbb{Q}}\left[H(S_{t_1}, S_{t_2}, \ldots, S_{t_n})\delta\left(\sigma^{-1}(\cdot, S_.)Y_. \widetilde{\beta}_.^{a,\widetilde{\sigma}} \delta_0(\cdot)\right)\right].$$

Moreover, if the process $\left(\beta_t^{\widetilde{\sigma}} \delta_0(z)\right)_{t\in[0,T]}$ belongs to $\mathbb{D}^{1,2}$, then

$$\begin{aligned}
\delta\left(\sigma^{-1}(\cdot, S_.)Y_. \widetilde{\beta}_.^{a,\widetilde{\sigma}} \delta_0(\cdot)\right) = \sum_{i=1}^{n} \Bigg\{ & \beta_{t_i}^{\widetilde{\sigma}} \delta_0(z) \int_{t_{i-1}}^{t_i} a(t)(\sigma^{-1}(t, S_{t-})Y_{t-}) dW_t \\
& - \int_{t_{i-1}}^{t_i} a(t)\left((D_{t,0}\beta_{t_i}^{\widetilde{\sigma}})\sigma^{-1}(t, S_{t-})Y_{t-}\right) dt \\
& - \int_{t_{i-1}}^{t_i} a(t)(\sigma^{-1}(t, S_{t-})Y_{t-}\beta_{t_{i-1}}^{\widetilde{\sigma}} \delta_0(z) dW_t \Bigg\}.
\end{aligned}$$

Proof Let H be a continuously differentiable function with bounded gradient. Then we can differentiate inside the expectation

$$\begin{aligned}
\frac{\partial \mathscr{C}^{\varepsilon,\widetilde{\sigma}}}{\partial \varepsilon}(x) &= \mathrm{E}_{\mathbb{Q}}\left[\sum_{i=1}^{n} \nabla_i H(S_{t_1}^{\varepsilon,\widetilde{\sigma}}, S_{t_2}^{\varepsilon,\widetilde{\sigma}}, \ldots, S_{t_n}^{\varepsilon,\widetilde{\sigma}}) \frac{\partial S_{t_i}^{\varepsilon,\widetilde{\sigma}}}{\partial \varepsilon}\right] \\
&= \mathrm{E}_{\mathbb{Q}}\left[\sum_{i=1}^{n} \nabla_i H(S_{t_1}^{\varepsilon,\widetilde{\sigma}}, S_{t_2}^{\varepsilon,\widetilde{\sigma}}, \ldots, S_{t_n}^{\varepsilon,\widetilde{\sigma}}) Z_{t_i}^{\varepsilon,\widetilde{\sigma}}\right].
\end{aligned}$$

Hence

$$\frac{\partial \mathscr{C}^{\varepsilon,\widetilde{\sigma}}}{\partial \varepsilon}(x)\bigg|_{\varepsilon=0} = \mathrm{E}_{\mathbb{Q}}\left[\sum_{i=1}^{n} \nabla_i H(S_{t_1}, S_{t_2}, \ldots, S_{t_n}) Z_{t_i}^{\widetilde{\sigma}}\right].$$

On the other hand we have

$$
\begin{aligned}
Z_{t_i}^{\widetilde{\sigma}} &= Y_{t_i} \beta_{t_i}^{\widetilde{\sigma}} \\
&= Y_{t_i} \sum_{j=1}^{i} (\beta_{t_j}^{\widetilde{\sigma}} - \beta_{t_{j-1}}^{\widetilde{\sigma}}) \\
&= Y_{t_i} \sum_{j=1}^{i} \int_{t_{j-1}}^{t_j} a(t)(\beta_{t_j}^{\widetilde{\sigma}} - \beta_{t_{j-1}}^{\widetilde{\sigma}}) dt \\
&= \int_{t_0}^{t_i} Y_{t_i} \widetilde{\beta}_t^{a,\widetilde{\sigma}} dt.
\end{aligned}
$$

From Proposition 2.3, we conclude that

$$
Z_{t_i}^{\widetilde{\sigma}} = \int_0^T D_{u,0} S_{t_i} \sigma^{-1}(u, S_{u-}) Y_{u-} \widetilde{\beta}_u^{a,\widetilde{\sigma}} du.
$$

Which implies that

$$
\begin{aligned}
\frac{\partial \mathscr{C}^{\varepsilon,\widetilde{\sigma}}}{\partial \varepsilon}(x) \bigg|_{\varepsilon=0} &= \mathrm{E}_{\mathbb{Q}} \left[\int_0^T \sum_{i=1}^{n} \nabla_i H(S_{t_1}, S_{t_2}, \ldots, S_{t_n}) D_{u,0} S_{t_i} \sigma^{-1}(u, S_{u-}) Y_{u-} \widetilde{\beta}_u^{a,\widetilde{\sigma}} du \right] \\
&= \mathrm{E}_{\mathbb{Q}} \left[\int_0^T D_{u,0} H(S_{t_1}, S_{t_2}, \ldots, S_{t_n}) \sigma^{-1}(u, S_{u-}) Y_{u-} \widetilde{\beta}_u^{a,\widetilde{\sigma}} du \right].
\end{aligned}
$$

Using the duality formula in Sect. 4.14 and taking into account the fact that $(\sigma^{-1}(t, S_t) Y_t \widetilde{\beta}_t^{a,\widetilde{\sigma}} \delta_0(z))_{(t,z)}$ belongs to $Dom(\delta)$, we reach

$$
\frac{\partial \mathscr{C}^{\varepsilon,\widetilde{\sigma}}}{\partial \varepsilon}(x) \bigg|_{\varepsilon=0} = \mathrm{E}_{\mathbb{Q}} \left[H(S_{t_1}, S_{t_2}, \ldots, S_{t_n}) \delta \left(\sigma^{-1}(\cdot, S_{\cdot}) Y_{\cdot} \widetilde{\beta}_{\cdot}^{a,\widetilde{\sigma}} \delta_0(\cdot) \right) \right].
$$

2.5 Variation in the Jump Amplitude

To derive a stochastic weight for the sensitivity with respect to the amplitude parameter φ we use the same technique as in the Proposition 2.6. To do this, we consider the perturbed process

$$
\begin{cases}
dS_t^{\varepsilon,\widetilde{\varphi}} = b(t, S_{t-}^{\varepsilon,\widetilde{\varphi}}) dt + \sigma(t, S_{t-}^{\varepsilon,\widetilde{\varphi}}) dW_t \\
\qquad\qquad + \int_{\mathbb{R}_0^d} (\varphi(t, S_{t-}^{\varepsilon,\widetilde{\varphi}}, z) + \varepsilon \widetilde{\varphi}(t, S_{t-}^{\varepsilon,\widetilde{\varphi}}, z)) \widetilde{N}(dt, dz), \\
S_0^{\varepsilon,\widetilde{\varphi}} = x
\end{cases}
$$

where $\varepsilon \in [-1, 1]$ and $\widetilde{\varphi} : \mathbb{R}^+ \times \mathbb{R}^d \times \mathbb{R}^d \longrightarrow \mathbb{R}^{d\times d}$ is continuously differentiable function with bounded first derivative in the second direction. The variation process with respect to the parameter ε becomes

$$\begin{cases} dZ_t^{\varepsilon,\widetilde{\varphi}} &= b'(t, S_{t-}^{\varepsilon,\widetilde{\varphi}})Z_{t-}^{\varepsilon,\widetilde{\varphi}}dt + \sum_{k=1}^d \sigma_k'(t, S_{t-}^{\varepsilon,\widetilde{\varphi}})Z_{t-}^{\varepsilon,\widetilde{\varphi}}dW_t^{(k)} \\ &\quad + \int_{\mathbb{R}_0^d}\left(\varphi'(t, S_{t-}^{\varepsilon,\widetilde{\varphi}}, z) + \varepsilon\widetilde{\varphi}'(t, S_{t-}^{\varepsilon,\widetilde{\varphi}}, z)\right) Z_{t-}^{\varepsilon,\widetilde{\varphi}}\widetilde{N}(dt, dz) \\ &\quad + \int_{\mathbb{R}_0^d}\widetilde{\varphi}(t, S_{t-}^{\varepsilon,\widetilde{\varphi}}, z)\widetilde{N}(dt, dz), \\ Z_0^{\varepsilon,\widetilde{\varphi}} &= 0. \end{cases}$$

We can also relate to this perturbed process the perturbed price $\mathscr{C}^{\varepsilon,\widetilde{\varphi}}(x)$ defined by

$$\mathscr{C}^{\varepsilon,\widetilde{\varphi}}(x) := \mathrm{E}_{\mathbb{Q}}\left[H(S_{t_1}^{\varepsilon,\widetilde{\varphi}}, S_{t_1}^{\varepsilon,\widetilde{\varphi}}, \ldots, S_{t_n}^{\varepsilon,\widetilde{\varphi}})\right].$$

Hence, the statement of the following proposition is practically identical to Proposition 2.8:

Proposition 2.9 *Assume that the diffusion matrix σ is uniformly elliptic and the process $(\sigma^{-1}(t, S_t)Y_t\widetilde{\beta}_t^{a,\widetilde{\varphi}}\delta_0(z))_{(t,z)} \in Dom(\delta)$, then we have for any $a \in \widetilde{T}_n$:*

$$Kappa = \frac{\partial \mathscr{C}^{\varepsilon,\widetilde{\varphi}}}{\partial \varepsilon}(x)\bigg|_{\varepsilon=0} = \mathrm{E}_{\mathbb{Q}}\left[H(S_{t_1}, S_{t_2}, \ldots, S_{t_n})\delta\left(\sigma^{-1}(\cdot, S_\cdot)Y_\cdot\widetilde{\beta}_\cdot^{a,\widetilde{\varphi}}\delta_0(\cdot)\right)\right].$$

Moreover, if the process $(\beta_t^{\widetilde{\varphi}}\delta_0(z))_{t\in[0,T]}$ belongs to $\mathbb{D}^{1,2}$, then

$$\begin{aligned} \delta\left(\sigma^{-1}(\cdot, S_\cdot)Y_\cdot\widetilde{\beta}_\cdot^{a,\widetilde{\varphi}}\delta_0(\cdot)\right) = \sum_{i=1}^n \Bigg\{ & \beta_{t_i}^{\widetilde{\varphi}}\delta_0(z)\int_{t_{i-1}}^{t_i} a(t)(\sigma^{-1}(t, S_{t-})Y_{t-})dW_t \\ & - \int_{t_{i-1}}^{t_i} a(t)\left((D_{t,0}\beta_{t_i}^{\widetilde{\varphi}})\sigma^{-1}(t, S_{t-})Y_{t-}\right)dt \\ & - \int_{t_{i-1}}^{t_i} a(t)(\sigma^{-1}(t, S_{t-})Y_{t-}\beta_{t_{i-1}}^{\widetilde{\varphi}}\delta_0(z))dW_t \Bigg\}. \end{aligned}$$

3 Numerical Experiments

In this section, we provide some simple examples to illustrate the results achieved in the previous section. In particular, we will look at time-inhomogeneous versions of the Merton model and the Bates model.

3.1 Examples

3.1.1 Time-Inhomogeneous Merton Model

We consider time-inhomogeneous versions of the Merton model when the riskless asset is governed by the equation:

$$dS_t^0 = S_t^0 r(t)dt, \quad S_0^0 = 1,$$

and the evolution of the risky asset is described by:

$$dS_t = S_{t-}dL_t, \quad S_0 = x,$$

where

$$L_t = \int_0^t b(u)du + \int_0^t \sigma(u)dW_u + \int_0^t \varphi(u)dX_u, \quad t \geq 0.$$

- $\{W_t, 0 \leq t \leq T\}$ is a standard Brownian motion.
- The process $\{X_t, 0 \leq t \leq T\}$ is defined by $X_t := \sum_{j=1}^{N_t} Z_j$ for all $t \in [0, T]$, such that $\{N_t, t \geq 0\}$ is a inhomogeneous Poisson process with intensity function $\lambda(t)$ and $(Z_n)_{n\geq 1}$ is a sequence of square integrable random variables which are i.i.d. (we set $\kappa := \mathrm{E}_{\mathbb{Q}}[Z_1]$).
- $\{W_t, t \geq 0\}$, $\{N_t, t \geq 0\}$ and $\{Z_n, n \geq 1\}$ are independent.
- r, b, σ and φ are deterministic functions.

We can write

$$\begin{aligned} L_t &= \int_0^t b(u)du + \int_0^t \sigma(u)dW_u + \int_0^t \int_{\mathbb{R}_0} \varphi(u)zJ_X(du, dz) \\ &= \int_0^t (b(u) + \kappa\varphi(u)\lambda(u))\, du + \int_0^t \sigma(u)dW_u + \int_0^t \int_{\mathbb{R}_0} \varphi(u)z\widetilde{J}_X(du, dz), \end{aligned}$$

where $J_X(du, dz)$ and $\widetilde{J}_X(du, dz)$ are, respectively, the jump measure and the compensated jump measure of the process X. By Itô's formula, we have for all $t \in [0, T]$:

$$\begin{aligned} \ln(S_t) = \ln(x) &+ \int_0^t \left(b(u) - \frac{1}{2}\sigma^2(u)\right) du \\ &+ \int_0^t \sigma(u)dW_u + \int_0^t \int_{\mathbb{R}_0} \ln(1 + \varphi(u)z) J_X(du, dz). \end{aligned}$$

Set $A_t = \exp(-\int_0^t r(u)du)$, we conclude that the process $(A_t S_t)_{t\in[0,T]}$ is a martingale if and only if the following condition is satisfied:

$$b(t) - r(t) + \kappa\varphi(t)\lambda(t) = 0 \;\; \forall\, t \in [0, T].$$

Hence, for all $t \in [0, T]$:

$$\begin{aligned}\ln(S_t) = \ln(x) &+ \int_0^t \left(r(u) - \frac{1}{2}\sigma^2(u) - \kappa\varphi(u)\lambda(u)\right) du \\ &+ \int_0^t \sigma(u) dW_u + \int_0^t \int_{\mathbb{R}_0} \ln(1 + \varphi(u)z) J_X(du, dz).\end{aligned}$$

The price of a contingent claim $H(S_T)$ is then expressed as

$$\mathscr{C}(x) = \mathrm{E}_{\mathbb{Q}}\left[A_T H(S_T)\right],$$

and for all $t \in [0, T]$, the processes Y_t, $Z_t^{\widetilde{\sigma}}$, $\beta_t^{\widetilde{\sigma}}$, $Z_t^{\widetilde{\varphi}}$ and $\beta_t^{\widetilde{\varphi}}$ are, respectively, given by

$$\begin{aligned}Y_t &= \frac{S_t}{x} \\ Z_t^{\widetilde{\sigma}} &= \left(\int_0^t \widetilde{\sigma}(u) dW_u - \int_0^t \widetilde{\sigma}(u)\sigma(u) du\right) S_t \\ \beta_t^{\widetilde{\sigma}} &= x \left(\int_0^t \widetilde{\sigma}(u) dW_u - \int_0^t \widetilde{\sigma}(u)\sigma(u) du\right) \\ Z_t^{\widetilde{\varphi}} &= \left(\int_0^t \int_{\mathbb{R}_0} \frac{\widetilde{\varphi}(u)z}{1 + \varphi(u)z} J_X(du, dz) - \int_0^t \kappa\widetilde{\varphi}(u)\lambda(u) du\right) S_t \\ \beta_t^{\widetilde{\varphi}} &= x \left(\int_0^t \int_{\mathbb{R}_0} \frac{\widetilde{\varphi}(u)z}{1 + \varphi(u)z} J_X(du, dz) - \int_0^t \kappa\widetilde{\varphi}(u)\lambda(u) du\right).\end{aligned}$$

By using the general formulae developed in the previous section, we are able to compute analytically the values of the different Greeks ($a(u) = \frac{1}{T}$):

$$\begin{aligned}\nabla_x \mathscr{C}(x) &= \mathrm{E}_{\mathbb{Q}}\left[A_T H(S_T) \int_0^T a(u) \left(\sigma^{-1}(u, S_{u-}) Y_{u-}\right) dW_u\right] \\ &= \mathrm{E}_{\mathbb{Q}}\left[A_T H(S_T) \int_0^T \frac{1}{xT\sigma(u)} dW_u\right]\end{aligned}$$

$$
\begin{aligned}
Rho_{\widetilde{r}} &= \mathrm{E}_{\mathbb{Q}}\left[A_T H(S_T)\int_0^T \left(\sigma^{-1}(t, S_{t-})\widetilde{r}(t, S_{t-})\right) dW_t\right] \\
&\quad - \mathrm{E}_{\mathbb{Q}}\left[\int_0^T \widetilde{r}(u)du A_T H(S_T)\right] \\
&= \mathrm{E}_{\mathbb{Q}}\left[A_T H(S_T)\left(\int_0^T \frac{\widetilde{r}(u)}{\sigma(u)} dW_u - \int_0^T \widetilde{r}(u)du\right)\right]
\end{aligned}
$$

$$
\begin{aligned}
Vega_{\widetilde{\sigma}} &= \mathrm{E}_{\mathbb{Q}}\left[A_T H(S_T)\int_0^T \sigma^{-1}(t, S_{t-})Y_{t-}\widetilde{\beta}^a_{t-} dW_t\right] \\
&= \mathrm{E}_{\mathbb{Q}}\left[A_T H(S_T)\left(\int_0^T a(t)\beta_T(\sigma^{-1}(t, S_{t-})Y_{t-})dW_t\right)\right] \\
&= \mathrm{E}_{\mathbb{Q}}\left[A_T H(S_T)\left(\int_0^T \frac{a(t)}{\sigma(t)}\left(\int_0^T \widetilde{\sigma}(u)(dW_u - \sigma(u)du)\right) dW_t\right)\right]
\end{aligned}
$$

$$
\begin{aligned}
Kappa_{\widetilde{\varphi}} &= \mathrm{E}_{\mathbb{Q}}\left[A_T H(S_T)\int_0^T \sigma^{-1}(t, S_{t-})Y_{t-}\widetilde{\beta}^a_{t-} dW_t\right] \\
&= \mathrm{E}_{\mathbb{Q}}\left[A_T H(S_T)\left(\int_0^T a(t)\beta_T(\sigma^{-1}(t, S_{t-})Y_{t-})dW_t\right)\right] \\
&= \mathrm{E}_{\mathbb{Q}}\left[A_T H(S_T)\left(\int_0^T \frac{a(t)}{\sigma(t)} dW_t\right)\right. \\
&\quad \times \left.\left(\int_0^T \int_{\mathbb{R}_0} \frac{\widetilde{\varphi}(u)z}{1+\varphi(u)z} J_X(du, dz) - \int_0^T \kappa\widetilde{\varphi}(u)\lambda(u)du\right)\right]
\end{aligned}
$$

For numerical simplicity we suppose that the coefficients $r > 0$, $\sigma > 0$ are real constants and $\varphi = 1$ such that $\ln(1 + Z_1) \sim \mathcal{N}(\mu, \delta^2)$ where $\mu \in \mathbb{R}$ and $\delta > 0$. The intensity function $\lambda(t)$ is exponentially decreasing given by $\lambda(t) = ae^{-bt}$ for all $t \in [0, T]$, where $a > 0$ and $b > 0$.

In this case we have $\kappa = \mathrm{E}[Z_1] = e^{\mu+\frac{\delta^2}{2}} - 1$ and the mean–value function of the Poisson process $\{N_t, t \geq 0\}$ is $m(t) = \int_0^t \lambda(s)ds = \frac{a}{b}\left(1 - e^{-bt}\right), \quad \forall\, t \in [0, T]$.

3.1.2 Binary Call Option

We consider the payoff of a digital call option of strike $K > 0$ and maturity T i.e. $H(S_T) = \mathbf{1}_{\{S_T \geq K\}}$, such that:

$$
S_T = x \exp\left\{\left(r - \frac{1}{2}\sigma^2\right)T - \frac{a\kappa}{b}(1 - e^{-bT}) + \sigma W_T + \sum_{j=1}^{N_T} \ln(1 + Z_j)\right\}.
$$

The price of a digital option is given by:

$$\mathscr{C}_{bin}^{M} := \mathscr{C}(x) = e^{-rT}\mathrm{E}_{\mathbb{Q}}[\mathbf{1}_{[K,+\infty[}(S_T)].$$

Delta: variation in the initial condition

- Delta computed from a derivation under expectation: By conditioning on the number of jumps, we can express the price as a weighted sum of Black–Scholes prices:

$$\mathscr{C}_{bin}^{M} = \sum_{n\geq 0} \frac{e^{-m(T)}(m(T))^n}{n!} \mathscr{C}_{bin}^{BS}(0, T, S_n, K, r, \sigma_n)$$

where $m(T) = \frac{a}{b}(1 - e^{-bT})$, $S_n = x\exp(n(\mu + \frac{\delta^2}{2}) - m(T)\kappa)$, $\sigma_n^2 = \sigma^2 + n\frac{\delta^2}{2}$ and $\mathscr{C}_{bin}^{BS}(0, T, S_n, K, r, \sigma_n)$ stands for the Black–Scholes price of a digital option.

$$\Delta_{bin}^{M} := \frac{\partial \mathscr{C}_{bin}^{M}}{\partial x} = \sum_{n\geq 0} \frac{e^{-m(T)}(m(T))^n}{n!} \frac{S_n}{x} \frac{\partial \mathscr{C}_{bin}^{BS}(0, T, S_n, K, r, \sigma_n)}{\partial S_n}.$$

Recall that

$$\mathscr{C}_{bin}^{BS}(0, T, S_n, K, r, \sigma_n) = e^{-rT}\mathscr{N}(d_{2,n})$$

and

$$\frac{\partial \mathscr{C}_{bin}^{BS}(0, T, S_n, K, r, \sigma_n)}{\partial S_n} = \frac{e^{-rT}}{S_n\sigma_n\sqrt{T}}\Phi(d_{2,n})$$

where $d_{1,n} = \frac{\ln(\frac{S_n}{K}) + (r + \frac{\sigma_n^2}{2})T}{\sigma_n\sqrt{T}}$, $d_{2,n} = d_{1,n} - \sigma_n\sqrt{T}$ and $\Phi(z) = \frac{1}{\sqrt{2\pi}}e^{\frac{-z^2}{2}}$. Consequently

$$\Delta_{bin}^{M} = \frac{e^{-(rT+m(T))}}{x\sqrt{T}} \sum_{n\geq 0} \frac{(m(T))^n}{n!} \frac{\Phi(d_{2,n})}{\sigma_n}.$$

- Finite difference approximation scheme of Delta:

$$\Delta_{bin}^{M,DF} = \frac{\partial}{\partial x}\mathrm{E}_{\mathbb{Q}}[e^{-rT}H(S_T^x)] \simeq \frac{\mathrm{E}_{\mathbb{Q}}[e^{-rT}H(S_T^{x+\varepsilon})] - \mathrm{E}_{\mathbb{Q}}[e^{-rT}H(S_T^{x-\varepsilon})]}{2\varepsilon}.$$

- Global Malliavin formula for Delta:
  The stochastic Malliavin weight for the delta is written:

$$\delta(\omega) = \int_0^T \frac{1}{T} \frac{S_t}{x\sigma S_t} dW_t = \frac{W_T}{x\sigma T}$$

where $\omega(t) = a(t)\frac{Y_t}{\sigma S_t}$ and $Y_t = \frac{S_t}{x}$ and $a(t) = \frac{1}{T}$

$$\Delta_{bin}^{M,Mal} = \mathrm{E}_{\mathbb{Q}}\left[e^{-rT} \mathbf{1}_{[K,+\infty]}(S_T) \frac{W_T}{x\sigma T} \right].$$

- Localized Malliavin formula for Delta:
  Empirical studies have shown that the theoretical estimators produced by the techniques of Malliavin are unbiased. We will adopt the localization technique introduced by Fournié et al. [9], which aims is to reduce the variance of the Monte–Carlo estimator for the sensitivities by localizing the integration by part formula around the singularity at K.
  Consider the decomposition:

$$H(S_T) = H_{\varepsilon,loc}(S_T) + H_{\varepsilon,reg}(S_T).$$

The regular component is defined by:

$$H_{\varepsilon,reg}(S_T) := G_\varepsilon(S_T - K).$$

where ε is a localization parameter and the localization function G_ε, that we propose, is given by:

$$G_\varepsilon(z) = \begin{cases} 0; & z \leq -\varepsilon \\ \frac{1}{2}\left(1 - \frac{z}{\varepsilon}\right)\left(1 + \frac{z}{\varepsilon}\right)^3; & -\varepsilon < z < 0 \\ 1 - \frac{1}{2}\left(1 + \frac{z}{\varepsilon}\right)\left(1 - \frac{z}{\varepsilon}\right)^3; & 0 \leq z < \varepsilon \\ 1; & z \geq \varepsilon. \end{cases}$$

Then

$$\begin{aligned} H_{\varepsilon,reg}(S_T) &= \frac{1}{2}\left(1 - \frac{S_T - K}{\varepsilon}\right)\left(1 + \frac{S_T - K}{\varepsilon}\right)^3 \mathbf{1}_{\{K-\varepsilon<S_T<K\}} \\ &+ \left(1 - \frac{1}{2}\left(1 + \frac{S_T - K}{\varepsilon}\right)\left(1 - \frac{S_T - K}{\varepsilon}\right)^3\right) \mathbf{1}_{\{K \leq S_T < K+\varepsilon\}} \\ &+ \mathbf{1}_{\{S_T \geq K+\varepsilon\}}. \end{aligned}$$

The localized component is given by:

$$H_{\varepsilon,loc}(S_T) = H(S_T) - H_{\varepsilon,reg}(S_T).$$

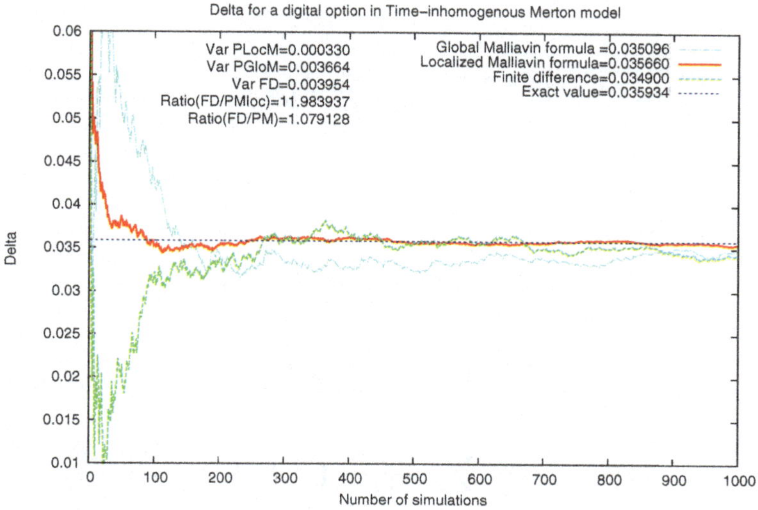

Fig. 1 Delta of a digital option computed by global, localized Malliavin like formula and finite difference. The parameters are $S_0 = 100$, $K = 100$, $\sigma = 0.10$, $T = 1$, $r = 0.02$, $\mu = -0.05$, $\delta = 0.01$, $\varphi = 1$, the intensity function λ is exponentially decreasing given by $\lambda(t) = ae^{-bt}$ for all $t \in [0, T]$, where $a = 1$ and $b = 1$

We find that the Delta computed by localized Malliavin formula:

$$\Delta_{LocMall} = e^{-rT}\mathrm{E}_{\mathbb{Q}}\left[H_{\varepsilon,loc}(S_T)\frac{W_T}{x\sigma T}\right] + e^{-rT}\mathrm{E}_{\mathbb{Q}}\left[H'_{\varepsilon,reg}(S_T)\frac{S_T}{x}\right].$$

In Fig. 1 we plot the delta for a digital option for a simplest time-inhomogeneous Merton model.

Furthermore, we have

$$Rho = e^{-rT}\mathrm{E}_{\mathbb{Q}}\left[\left(\frac{W_T}{\sigma} - T\right)\mathbf{1}_{\{S_T \geq K\}}\right]$$

$$Vega = e^{-rT}\mathrm{E}_{\mathbb{Q}}\left[\left(\frac{W_T^2 - \sigma T W_T - T}{\sigma T}\right)\mathbf{1}_{\{S_T \geq K\}}\right]$$

$$Kappa = e^{-rT}\mathrm{E}_{\mathbb{Q}}\left[\left(\sum_{j=1}^{N_T}\frac{Z_j}{1+\varphi Z_j} - \kappa\frac{a}{b}(-e^{-bT}+1)\right)\frac{W_T}{\sigma T}\mathbf{1}_{\{S_T \geq K\}}\right].$$

Rho: variation in the drift coefficient

- Rho computed from a derivation under expectation: Recall that

$$\mathscr{C}_{bin}^{BS}(0, T, S_n, K, r, \sigma_n) = e^{-rT}\mathscr{N}(d_{2,n})$$

and

$$\frac{\partial \mathscr{C}_{bin}^{BS}(0,T,S_n,K,r,\sigma_n)}{\partial r} = -Te^{-rT}\mathscr{N}(d_{2,n}) + \frac{\sqrt{T}e^{-rT}}{\sigma_n}\Phi(d_{2,n})$$

$$\begin{aligned} Rho_{bin}^{M} &:= \frac{\partial \mathscr{C}_{bin}^{M}}{\partial r} \\ &= \sum_{n\geq 0} \frac{e^{-m(T)}(m(T))^n}{n!} \frac{\partial \mathscr{C}_{bin}^{BS}(0,T,S_n,K,r,\sigma_n)}{\partial r} \\ &= \sum_{n\geq 0} \frac{e^{-m(T)}(m(T))^n}{n!}\left(-Te^{-rT}\mathscr{N}(d_{2,n}) + \frac{\sqrt{T}e^{-rT}}{\sigma_n}\Phi(d_{2,n})\right) \\ &= Te^{-(rT+m(T))}\sum_{n\geq 0}\frac{(m(T))^n}{n!}\left(-\mathscr{N}(d_{2,n}) + \frac{\Phi(d_{2,n})}{\sqrt{T}\sigma_n}\right). \end{aligned}$$

- Finite Difference Approximation scheme of Rho:

$$Rho_{FD} := \frac{\partial}{\partial r}\mathrm{E}_{\mathbb{Q}}[e^{-rT}H(S_T)] \simeq \frac{\mathrm{E}_{\mathbb{Q}}[e^{-(r+\varepsilon)T}H(S_T^{r+\varepsilon})] - \mathrm{E}_{\mathbb{Q}}[e^{-(r-\varepsilon)T}H(S_T^{r-\varepsilon})]}{2\varepsilon}.$$

- Global Malliavin formula for Rho:

$$Rho_{GMall} = e^{-rT}\mathrm{E}_{\mathbb{Q}}\left[\left(\frac{W_T}{\sigma} - T\right)\mathbf{1}_{\{S_T\geq K\}}\right].$$

- Localized Malliavin formula for Rho:

$$\begin{aligned} Rho_{LocMall} &= e^{-rT}\mathrm{E}_{\mathbb{Q}}\left[H_{\varepsilon,loc}(S_T)\left(\frac{W_T}{\sigma} - T\right)\right] \\ &\quad + e^{-rT}\mathrm{E}_{\mathbb{Q}}\left[H'_{\varepsilon,reg}(S_T)TS_T\right] - Te^{-rT}\mathrm{E}_{\mathbb{Q}}\left[H'_{\varepsilon,reg}(S_T)\right]. \end{aligned}$$

In Fig. 2 we plot the Rho for a digital option for a simplest time-inhomogeneous Merton model.

Vega: variation in the diffusion coefficient

- Vega computed from a derivation under expectation:

$$\begin{aligned} Vega_{bin}^{M} &:= \frac{\partial \mathscr{C}_{bin}^{M}}{\partial \sigma} \\ &= \sum_{n\geq 0}\frac{e^{-m(T)}(m(T))^n}{n!}\frac{\partial \sigma_n}{\partial \sigma}\frac{\partial \mathscr{C}_{bin}^{BS}(0,T,S_n,K,r,\sigma_n)}{\partial \sigma_n} \end{aligned}$$

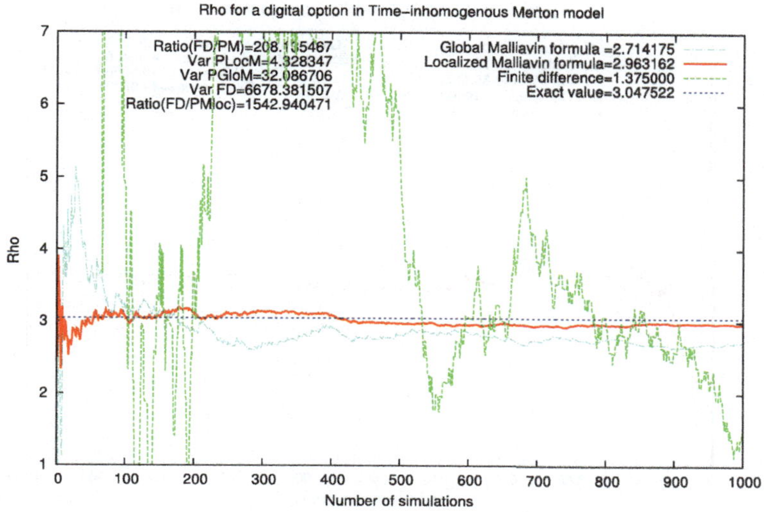

Fig. 2 Rho of a digital option computed by global, localized Malliavin like formula and finite difference. The parameters are $S_0 = 100$, $K = 100$, $\sigma = 0.1$, $T = 1$, $r = 0.03$, $\mu = -0.05$, $\delta = 0.01$, $\varphi = 1$, the intensity function λ is exponentially decreasing given by $\lambda(t) = ae^{-bt}$ for all $t \in [0, T]$, where $a = 1$ and $b = 1$

$$= \sum_{n\geq 0} \frac{e^{-m(T)}(m(T))^n}{n!} \frac{\sigma}{\sigma_n}(-e^{-rT})(\sqrt{T} + \frac{d_{2,n}}{\sigma_n})\Phi(d_{2,n})$$

$$= -\sigma e^{-(rT+m(T))} \sum_{n\geq 0} \frac{(m(T))^n}{n!} \left(\frac{\sigma_n\sqrt{T} + d_{2,n}}{\sigma_n^2} \right) \Phi(d_{2,n}).$$

- Finite Difference Approximation scheme of Vega:

$$Vega_{FD} := \frac{\partial}{\partial\sigma} \mathrm{E}_{\mathbb{Q}}[e^{-rT} H(S_T^{\sigma})] \simeq e^{-rT} \frac{\mathrm{E}_{\mathbb{Q}}[H(S_T^{\sigma+\varepsilon})] - \mathrm{E}_{\mathbb{Q}}[H(S_T^{\sigma-\varepsilon})]}{2\varepsilon}.$$

- Global Malliavin formula for Vega:

$$Vega_{GMall} = e^{-rT} \mathrm{E}_{\mathbb{Q}} \left[\left(\frac{W_T^2 - \sigma T W_T - T}{\sigma T} \right) \mathbf{1}_{\{S_T \geq K\}} \right].$$

- Localized Malliavin formula for Vega:

$$\begin{aligned} Vega_{LocMall} &= e^{-rT} \mathrm{E}_{\mathbb{Q}} \left[H_{\varepsilon,loc}(S_T) \left(\frac{W_T^2 - \sigma T W_T - T}{\sigma T} \right) \right] \\ &+ e^{-rT} \mathrm{E}_{\mathbb{Q}} \left[H'_{\varepsilon,reg}(S_T) \left(W_T - \sigma T \right) S_T \right]. \end{aligned}$$

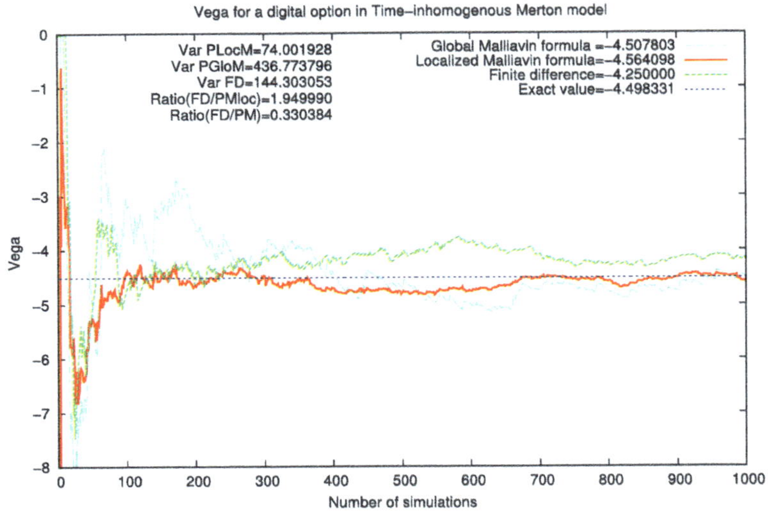

Fig. 3 Vega of a digital option computed by global, localized Malliavin like formula and finite difference. The parameters are $S_0 = 100$, $K = 100$, $r = 0.02$, $\sigma = 0.20$, $T = 1$, $r = 0.05$, $\mu = -0.05$, $\delta = 0.01$, $\varphi = 1$, the intensity function λ is exponentially decreasing given by $\lambda(t) = ae^{-bt}$ for all $t \in [0, T]$, where $a = 1$ and $b = 1$

In Fig. 3 we plot the Vega for a digital option for a simplest time-inhomogeneous Merton model.

Alpha: variation in the jump amplitude

- Alpha computed from a derivation under expectation:

$$\begin{aligned} Alpha^M_{bin} &:= \frac{\partial \mathscr{C}^M_{bin}}{\partial \varphi} \\ &= \sum_{n\geq 0} \frac{e^{-m(T)}(m(T))^n}{n!} \frac{\partial S_n}{\partial \varphi} \frac{\partial \mathscr{C}^{BS}_{bin}(0, T, S_n, K, r, \sigma_n)}{\partial S_n} \\ &= \sum_{n\geq 0} \frac{e^{-m(T)}(m(T))^n}{n!} \frac{m(T)\kappa S_n}{\varphi} \frac{\partial \mathscr{C}^{BS}_{bin}(0, T, S_n, K, r, \sigma_n)}{\partial S_n} \\ &= \frac{\kappa e^{-(rT+m(T))}}{\varphi\sqrt{T}} \sum_{n\geq 0} \frac{(m(T))^{n+1}}{n!} \frac{\Phi(d_{2,n})}{\sigma_n}. \end{aligned}$$

- Finite Difference Approximation scheme of Alpha:

$$Alpha_{FD} := \frac{\partial}{\partial \varphi} \mathrm{E}_{\mathbb{Q}}[e^{-rT} H(S^{\varphi}_T)] \simeq e^{-rT} \frac{\mathrm{E}_{\mathbb{Q}}[H(S^{\varphi+\varepsilon}_T)] - \mathrm{E}_{\mathbb{Q}}[H(S^{\varphi-\varepsilon}_T)]}{2\varepsilon}.$$

- Global Malliavin formula for Alpha:

$$Alpha_{GMall} = e^{-rT}\mathrm{E}_{\mathbb{Q}}\left[\left(\sum_{j=1}^{N_T}\frac{Z_j}{1+\varphi Z_j} - \kappa\frac{a}{b}(-e^{-bT}+1)\right)\frac{W_T}{\sigma T}\mathbf{1}_{\{S_T\geq K\}}\right].$$

- Localized Malliavin formula for Alpha:

$$Alpha_{LocMall} = e^{-rT}\mathrm{E}_{\mathbb{Q}}\left[H_{\varepsilon,loc}(S_T)\left(\sum_{j=1}^{N_T}\frac{Z_j}{1+\varphi Z_j} - \kappa\frac{a}{b}(-e^{-bT}+1)\right)\frac{W_T}{\sigma T}\right]$$
$$+e^{-rT}\mathrm{E}_{\mathbb{Q}}\left[H'_{\varepsilon,reg}(S_T)\left(\sum_{j=1}^{N_T}\frac{Z_j}{1+\varphi Z_j} - \kappa\frac{a}{b}(-e^{-bT}+1)\right)S_T\right].$$

In Fig. 4 we plot the sensitivity with respect to the jump size parameter φ for a digital option for a simplest time-inhomogeneous Merton model.

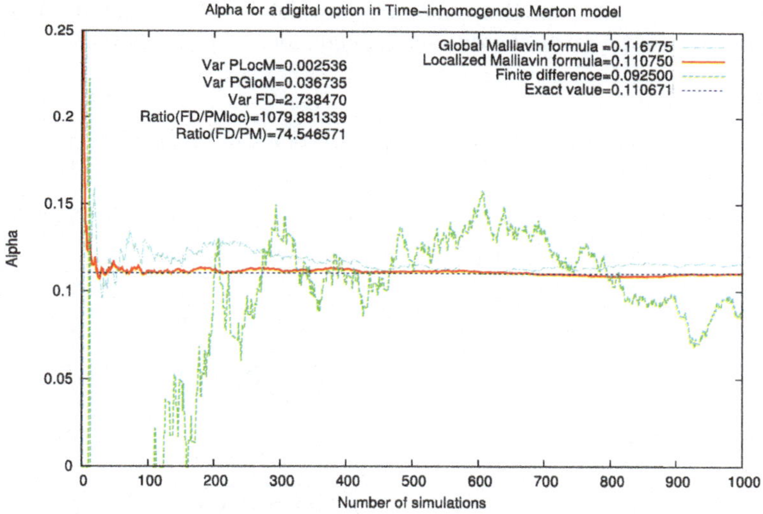

Fig. 4 Alpha of a digital option computed by global, localized Malliavin like formula and finite difference. The parameters are $S_0 = 100$, $K = 100$, $\sigma = 0.20$, $T = 1$, $r = 0.02$, $\mu = -0.05$, $\delta = 0.01$, $\varphi = 1$, the intensity function λ is exponentially decreasing given by $\lambda(t) = ae^{-bt}$ for all $t \in [0, T]$, where $a = 1$ and $b = 1$

3.1.3 Time-Inhomogeneous Bates Model:

We consider the solution of the stochastic differential equation:

$$\begin{cases} dS_t^1 = rS_{t-}^1 dt + \sqrt{V_t} S_{t-}^1 dW_t^1 + S_{t-}^1 \int_{\mathbb{R}_0} (e^z - 1)\widetilde{N}(dt, dz), \quad S_0^1 = x_0, \\ dV_t = \kappa(\theta - V_t)dt + \sigma\sqrt{V_t} dB_t, \quad V_0 = v_0, \\ \langle W^1, B \rangle_t = \rho t, \end{cases}$$

where $(W_t^1, B_t)_{t\in[0,T]}$ is a two–dimensional correlated Brownian motion with correlation parameter $\rho \in]-1, 1[$. The stochastic process (S_t^1) is the underling price process and (V_t) is the square of the volatility process which follows a CIR[2] process with an initial value $v_0 > 0$, with long–run mean θ, and rate of reversion κ, σ is referred to as the volatility of volatility.
For all $t \in [0, T]$, we define

$$W_t^2 := \frac{1}{\sqrt{1-\rho^2}} \left(B_t - \rho W_t^1\right).$$

The process $(W_t^2)_{t\in[0,T]}$ is a Brownian motion which is independent of $(W_t^1)_{t\in[0,T]}$. Then, the system of stochastic differential equations can be rewritten in a matrix form

$$dS_t = b(t, S_{t-})dt + \sigma(t, S_{t-})dW_t + \int_{\mathbb{R}_0} \varphi(t, S_{t-}, z)\widetilde{N}(dt, dz), \; S_0 = (x_0, v_0)$$

where $S_t = (S_t^1, V_t)$, $W_t^* = (W_t^1, W_t^2)^*$, $b^*(t, S_{t-}) = (rS_{t-}^1, \kappa(\theta - V_t))^*$, $\varphi^*(t, S_{t-}, z) = ((e^z - 1)S_{t-}^1, 0)^*$ and

$$\sigma(t, S_{t-}) = \begin{pmatrix} \sqrt{V_t} S_{t-}^1 & 0 \\ \rho\sigma\sqrt{V_t} & \sigma\sqrt{1-\rho^2}\sqrt{V_t} \end{pmatrix}.$$

The inverse of σ is

$$\sigma^{-1}(t, S_{t-}) = \frac{1}{\sigma\sqrt{1-\rho^2} S_{t-}^1 V_t} \begin{pmatrix} \sigma\sqrt{1-\rho^2}\sqrt{V_t} & 0 \\ -\rho\sigma\sqrt{V_t} & \sqrt{V_t} S_{t-}^1 \end{pmatrix}.$$

The price of the contingent claim in this setting is expressed as:

$$\mathscr{C} = \mathrm{E}_{\mathbb{Q}}\left[e^{-rT} H(S_t)\right].$$

[2]Cox, Ingersoll and Ross model. See [4].

Note that by Itô's formula we have for all $t \in [0, T]$

$$\ln(S_t^1) = \int_0^t \left(r - \frac{1}{2}V_u\right) du + \int_0^t \int_{\mathbb{R}_0} \left[z - (e^z - 1)\right] \nu_u(dz)du + \int_0^t \sqrt{V_u} dW_u^1 + \int_0^t \int_{\mathbb{R}_0} z\widetilde{N}(du, dz).$$

The Rho

In the drift—perturbed process $(S_t^\varepsilon)_t$ which is a solution of the stochastic differential equation (10), we take $\widetilde{b}^*(t, x) = (x_1, 0)^*$ and we get

$$(\sigma^{-1}(t, S_{t-})\widetilde{b}(t, S_{t-}))^* = \left(\frac{1}{\sqrt{V_t}}, \frac{-\rho}{\sqrt{1-\rho^2}\sqrt{V_t}}\right).$$

From Proposition 2.6, we have

$$Rho = e^{-rT}\mathrm{E}_{\mathbb{Q}}\left[H(S_t)\left(\int_0^T \frac{dW_t^1}{\sqrt{V_t}} - \frac{\rho}{\sqrt{1-\rho^2}}\int_0^T \frac{dW_t^2}{\sqrt{V_t}}\right)\right] - Te^{-rT}\mathrm{E}_{\mathbb{Q}}\left[H(S_t)\right].$$

The Delta

The first variation process is given by

$$\begin{cases} dY_t = b'(t, S_{t-})Y_{t-}dt + \sigma_1'(t, S_{t-})Y_{t-}dW_t^1 \\ \qquad + \sigma_2'(t, S_{t-})Y_{t-}dW_t^2 + \int_{\mathbb{R}_0} \varphi'(t, S_{t-}, z)Y_{t-}\widetilde{N}(dt, dz), \\ Y_0 = I_2 \end{cases}$$

where

$$b'(t, S_{t-}) = \begin{pmatrix} r & 0 \\ 0 & -\kappa \end{pmatrix}, \quad \varphi'(t, S_{t-}, z) = \begin{pmatrix} (e^z - 1) & 0 \\ 0 & 0 \end{pmatrix},$$

$$\sigma_1'(t, S_{t-}) = \begin{pmatrix} \sqrt{V_t} & \frac{S_{t-}^1}{2\sqrt{V_t}} \\ 0 & \frac{\sigma\rho}{2\sqrt{V_t}} \end{pmatrix} \text{ and } \sigma_2'(t, S_{t-}) = \begin{pmatrix} 0 & 0 \\ 0 & \frac{\sigma\sqrt{1-\rho^2}}{2\sqrt{V_t}} \end{pmatrix},$$

$$\left(\sigma^{-1}(t, S_{t-})Y_{t-}\right)^* = \begin{pmatrix} \frac{Y_{t-}^{1,1}}{S_{t-}^1\sqrt{V_t}} & \frac{-\rho}{\sqrt{1-\rho^2}}\frac{Y_{t-}^{1,1}}{S_{t-}^1\sqrt{V_t}} \\ \frac{Y_{t-}^{2,1}}{S_{t-}^1\sqrt{V_t}} & \frac{1}{\sqrt{1-\rho^2}\sqrt{V_t}}\left(\frac{-\rho Y_{t-}^{1,2}}{S_{t-}^1} + \frac{Y_{t-}^{2,2}}{\sigma}\right) \end{pmatrix}.$$

By Proposition 2.5 we conclude that

$$
\begin{aligned}
Delta &:= \frac{\partial \mathscr{C}}{\partial x_0} \\
&= e^{-rT}\mathrm{E}_{\mathbb{Q}}\left[H(S_t)\left(\int_0^T a(t)\frac{Y_{t-}^{1,1}}{S_{t-}^1\sqrt{V_t}}dW_t^1 - \int_0^T a(t)\frac{-\rho}{\sqrt{1-\rho^2}}\frac{Y_{t-}^{1,1}}{S_{t-}^1\sqrt{V_t}}dW_t^2\right)\right].
\end{aligned}
$$

Since $Y_{t-}^{1,1} = \frac{S_{t-}^1}{x_0}$ and if we take $a(t) = \frac{1}{T}$, we get

$$
Delta = \frac{e^{-rT}}{x_0 T}\mathrm{E}_{\mathbb{Q}}\left[H(S_t)\left(\int_0^T \frac{dW_t^1}{\sqrt{V_t}} - \frac{\rho}{\sqrt{1-\rho^2}}\int_0^T \frac{dW_t^2}{\sqrt{V_t}}\right)\right].
$$

The Vega

We perturb the original diffusion matrix with $\widetilde{\sigma}$ to get the perturbed process given by (11) such that

$$
\widetilde{\sigma}(t,x) = \begin{pmatrix} x_1 & 0 \\ 0 & 0 \end{pmatrix}.
$$

For all $t \in [0,T]$, the processes $Z_t^{\widetilde{\sigma}}$ and $\beta_t^{\widetilde{\sigma}}$ are, respectively, given by

$$
\begin{aligned}
Z_t^{1,\widetilde{\sigma}} &= \left(W_t^1 - \int_0^t \sqrt{V_u}du\right)S_t,\ Z_t^{2,\widetilde{\sigma}} = 0 \\
\beta_t^{1,\widetilde{\sigma}} &= x_0\left(W_t^1 - \int_0^t \sqrt{V_u}du\right),\ \beta_t^{2,\widetilde{\sigma}} = 0.
\end{aligned}
$$

Using the chain rule (Proposition 4.12) on a sequence of continuously differentiable functions with bounded derivatives approximating $\sqrt{V_u}$, together with Proposition 2.3 we obtain

$$
\begin{aligned}
D_{t,0}\beta_T^{1,\widetilde{\sigma}} &= x_0\left((1,0)^* - \int_0^T \frac{1}{2\sqrt{V_u}}D_{t,0}\sqrt{V_u}du\right) \\
&= x_0\left((1,0) - \frac{\sigma}{2}\int_t^T \frac{\sqrt{V_t}}{\sqrt{V_u}}\frac{Y_u^{2,2}}{Y_t^{2,2}}\left(\rho,\sqrt{1-\rho^2}\right)du\right).
\end{aligned}
$$

Thus

$$
Tr\left((D_{t,0}\beta_T)\sigma^{-1}(t,S_{t-})Y_{t-}\right) = \frac{1}{\sqrt{V_t}}.
$$

Then

$$\delta\left(\sigma^{-1}(\cdot, S_\cdot)Y_\cdot\widetilde{\beta}^a_\cdot\delta_0(\cdot)\right) = \beta_T^{\widetilde{\sigma}*}\int_0^T a(t)(\sigma^{-1}(t, S_{t-})Y_{t-})^* dW_t$$
$$-\int_0^T a(t)Tr\left((D_{t,0}\beta_T)\sigma^{-1}(t, S_{t-})Y_{t-}\right)dt$$
$$= \left(W_T^1 - \int_0^T \sqrt{V_u}du\right)$$
$$\times\left(\int_0^T \frac{a(t)}{\sqrt{V_t}}dW_t^1 - \frac{\rho}{\sqrt{1-\rho^2}}\int_0^T \frac{a(t)}{\sqrt{V_t}}dW_t^2\right)$$
$$-\int_0^T \frac{a(t)}{\sqrt{V_t}}dt.$$

Consequently,

$$Vega_{\widetilde{\sigma}} = \frac{e^{-rT}}{T}\mathrm{E}_{\mathbb{Q}}\left[H(S_T)\left(\left(W_T^1 - \int_0^T \sqrt{V_u}du\right)\right.\right.$$
$$\left.\left.\times\left(\int_0^T \frac{dW_t^1}{\sqrt{V_t}} - \frac{\rho}{\sqrt{1-\rho^2}}\int_0^T \frac{dW_t^2}{\sqrt{V_t}}\right) - \int_0^T \frac{dt}{\sqrt{V_t}}\right)\right].$$

The alpha

We consider the perturbed process

$$\begin{cases} dS_t^\varepsilon = b(t, S_{t-}^\varepsilon)dt + \sigma(t, S_{t-}^\varepsilon)dW_t \\ \qquad\quad + \int_{\mathbb{R}_0}(\varphi(t, S_{t-}^\varepsilon, z) + \varepsilon\widetilde{\varphi}(t, S_{t-}^\varepsilon, z))\widetilde{N}(dt, dz), \\ S_0^\varepsilon = x, \end{cases}$$

with

$$\widetilde{\varphi}(t, x, z) = \begin{pmatrix} x_1 \\ 0 \end{pmatrix}.$$

For all $t \in [0, T]$, the processes $Z_t^{\widetilde{\varphi}}$ and $\beta_t^{\widetilde{\varphi}}$ defined above are, respectively, given by

$$Z_t^{1,\widetilde{\varphi}} = \left(\int_0^t \int_{\mathbb{R}_0} e^{-z}\widetilde{N}(du, dz) - \int_0^t \int_{\mathbb{R}_0}(1 - e^{-z})\nu_u(dz)du\right)S_t, \quad Z_t^{2,\widetilde{\varphi}} = 0$$
$$\beta_t^{1,\widetilde{\varphi}} = x_0\left(\int_0^t \int_{\mathbb{R}_0} e^{-z}\widetilde{N}(du, dz) - \int_0^t \int_{\mathbb{R}_0}(1 - e^{-z})\nu_u(dz)du\right), \quad \beta_t^{2,\widetilde{\varphi}} = 0.$$

Then

$$\delta\left(\sigma^{-1}(\cdot,S_\cdot)Y_\cdot\widetilde{\beta}^a_\cdot\delta_0(\cdot)\right)=\beta_T^{\widetilde{\varphi}*}\int_0^T a(t)(\sigma^{-1}(t,S_{t-})Y_{t-})^* dW_t$$
$$-\int_0^T a(t)Tr\left((D_{t,0}\beta_T)\sigma^{-1}(t,S_{t-})Y_{t-}\right)dt$$
$$=\left(\int_0^T\int_{\mathbb{R}_0} e^{-z}\widetilde{N}(du,dz)-\int_0^T\int_{\mathbb{R}_0}(1-e^{-z})\nu_u(dz)du\right)$$
$$\times\left(\int_0^T\frac{a(t)}{\sqrt{V_t}}dW_t^1-\frac{\rho}{\sqrt{1-\rho^2}}\int_0^T\frac{a(t)}{\sqrt{V_t}}dW_t^2\right).$$

Consequently

$$Alpha_{\widetilde{\varphi}}=\frac{e^{-rT}}{T}\mathrm{E}_{\mathbb{Q}}\left[H(S_T)\left(\int_0^T\int_{\mathbb{R}_0} e^{-z}\widetilde{N}(du,dz)-\int_0^T\int_{\mathbb{R}_0}(1-e^{-z})\nu_u(dz)du\right)\right.$$
$$\left.\times\left(\int_0^T\frac{dW_t^1}{\sqrt{V_t}}-\frac{\rho}{\sqrt{1-\rho^2}}\int_0^T\frac{dW_t^2}{\sqrt{V_t}}\right)\right].$$

4 Malliavin Calculus for Square Integrable Additive Processes

4.1 Additive Processes

Definition 4.1 *(see Cont* [3], *Definition 14.1)* A stochastic process $(S_t)_{t\geq 0}$ on $\mathbb{R}^d$ is called an additive process if it is càdlàg, satisfies $S_0=0$ and has the following properties:

1. Independent increments: for every increasing sequence of times $t_0,\ldots,t_n$, the random variables $S_{t_0}, S_{t_1}-S_{t_0},\ldots,S_{t_n}-S_{t_{n-1}}$ are independent.
2. Stochastic continuity: $\forall\,\varepsilon>0$ and $\forall\,t\geq 0$, $\lim_{h\to 0}\mathbb{P}[|S_{t+h}-S_t|\geq\varepsilon]=0$.

Theorem 4.2 (see Sato [15], Theorems 9.1–9.8) *Let $(S_t)_{t\geq 0}$ be an additive process on $\mathbb{R}^d$. Then S_t has an infinitely divisible distribution for all t. The law of $(S_t)_{t\geq 0}$ is uniquely determined by its spot characteristics $(A_t,\mu_t,\Gamma_t)_{t\geq 0}$:*

$$\mathrm{E}[\exp(iuS_t)]=\exp(\psi_t(u))$$

where

$$\psi_t(u)=-\frac{1}{2}u\cdot A_t u+iu\cdot\Gamma_t+\int_{\mathbb{R}^d}(e^{iu\cdot z}-1-iu\cdot z\mathbf{1}_{\{|z|\leq 1\}})\mu_t(dz).$$

The spot characteristics $(A_t, \mu_t, \Gamma_t)_{t\geq 0}$ *satisfy the following conditions*

1. *For all* t, A_t *is a positive definite* $d \times d$ *matrix and* μ_t *is a positive measure on* $\mathbb{R}^d$ *satisfying* $\mu_t(0) = 0$ *and* $\int_{\mathbb{R}_0^d}(|z|^2 \wedge 1)\mu_t(dz) < \infty$.
2. *Positiveness:* $A_0 = 0$, $\mu_0 = 0$, $\Gamma_0 = 0$ *and for all* s, t *such that* $s \leq t$, $A_t - A_s$ *is a positive definite* $d \times d$ *matrix and* $\mu_t(B) \geq \mu_s(B)$ *for all measurable sets* $B \in \mathscr{B}(\mathbb{R}^d)$.
3. *Continuity: if* $s \longrightarrow t$ *then* $A_s \longrightarrow A_t$, $\Gamma_s \longrightarrow \Gamma_t$ *and* $\mu_s(B) \longrightarrow \mu_t(B)$ *for all* $B \in \mathscr{B}(\mathbb{R}^d)$ *such that* $B \subset \{z : |z| \geq \varepsilon\}$ *for some* $\varepsilon > 0$.

Conversely, for a family of $(A_t, \mu_t, \Gamma_t)_{t\geq 0}$ *satisfying the conditions* (1), (2) *and* (3) *above there exists an additive process* $(S_t)_{t\geq 0}$ *with* $(A_t, \mu_t, \Gamma_t)_{t\geq 0}$ *as spot characteristics.*

Example 1 We consider a class of spot characteristics $(A_t, \mu_t, \Gamma_t)_{t\geq 0}$ constructed in the following way:

- A continuous matrix valued function $\sigma : [0, T] \longrightarrow M_{d\times d}(\mathbb{R})$ such that σ_t is symmetric for all $t \in [0, T]$ and verifies $\int_0^T \sigma_t^2 dt < \infty$.
- A family $(\nu_t)_{t\in[0,T]}$ of Lévy measures verifying $\int_0^T \left(\int_{\mathbb{R}_0^d}(|z|^2 \wedge 1)\nu_t(dz)\right) dt < \infty$.
- A deterministic function with finite variation $\gamma : [0, T] \longrightarrow \mathbb{R}^d$ (e.g., a piecewise continuous function).

Then the spot characteristics $(A_t, \mu_t, \Gamma_t)_{t\geq 0}$ defined by

$$A_t = \int_0^t \sigma_s^2 ds$$

$$\mu_t = \int_0^t \nu_s ds$$

$$\Gamma_t = \int_0^t \gamma_s ds$$

satisfy the conditions 1, 2, 3 and therefore define a unique additive process $(S_t)_{t\geq 0}$ with spot characteristics $(A_t, \mu_t, \Gamma_t)_{t\in[0,T]}$. The triplet $(\sigma_t^2, \nu_t, \gamma_t)_{t\in[0,T]}$ are called local characteristics of the additive process.

Remark 4.3 Not all additive processes can be parameterized in this way, but we will assume this parametrization in terms of local characteristics in the rest of this paper. In particular, the assumptions above on the local characteristics implies that the process $(S_t)_{t\geq 0}$ is a semimartingale which will allow us to apply the Itô formula.

The local characteristics of an additive process enable us to describe the structure of its sample paths: the positions and sizes of jumps of $(S_t)_{t\geq 0}$ are described by a Poisson random measure on $[0, T] \times \mathbb{R}^d$

$$J_S(\omega, \cdot) = \sum_{0\leq t\leq T;\, \Delta S_t \neq 0} \delta_{(t, \Delta S_t)}$$

with (time-inhomogeneous) intensity given by $\nu_t(dz)dt$:

$$\mathrm{E}[J_S([t_1,t_2]\times B)] = \mu_T([t_1,t_2]\times B) = \int_{t_1}^{t_2} \nu_s(B)ds.$$

The compensated Poisson random measure can therefore be defined by:

$$\widetilde{J}_S(\omega,dt,dz) = J_S(\omega,dt,dz) - \nu_t(dz)dt.$$

4.2 Isonormal Lévy Process (ILP)

Let μ and ν are σ–finite measures without atoms on the measurable spaces $(\mathrm{T},\mathscr{A})$ and $(\mathrm{T}\times \mathrm{X}_0,\mathscr{B})$ respectively.
Define a new measure

$$\pi(dt,dz) := \mu(dt)\delta_\Delta(dz) + \nu(dt,dz) \tag{11}$$

on a measurable space $(\mathrm{T}\times\mathrm{X},\mathscr{G})$, where $\mathrm{X}=\mathrm{X}_0\cup\Delta$, $\mathscr{G}=\sigma(\mathscr{A}\times\Delta,\mathscr{B})$ and $\delta_\Delta(dz)$ is the measure which gives mass one to the point Δ.

We assume that the Hilbert space $\mathscr{H}=\mathrm{L}^2(\mathrm{T}\times\mathrm{X},\mathscr{G},\pi)$ is separable.

Definition 4.4 We say that a stochastic process $\mathrm{L}=\{\mathrm{L}(h), h\in\mathscr{H}\}$ defined in a complete probability space $(\Omega,\mathscr{F},P)$ is an isonormal Lévy process (or Lévy process on $\mathscr{H}$) if the following conditions are satisfied:

1. The mapping $h \longrightarrow L(h)$ is linear.
2. $\mathrm{E}[e^{ixL(h)}] = \exp(\Psi(x,h))$, where

$$\Psi(x,h) = \int_{\mathrm{T}\times\mathrm{X}} \left((e^{ixh(t,z)} - 1 - ixh(t,z))\mathbf{1}_{\mathrm{X}_0}(z) - \frac{1}{2}x^2h^2(t,z)\mathbf{1}_\Delta(z)\right)\pi(dt,dz).$$

4.3 Generalized Orthogonal Polynomials (GOP)

Denote by $\overline{x}=(x_1,x_2,\ldots,x_n,\ldots)$ a sequence of real numbers. Define a function $F(z,\overline{x})$ by

$$F(z,\overline{x}) = \exp\left(\sum_{k=1}^{\infty}\frac{(-1)^{k+1}}{k}x_k\,z^k\right). \tag{12}$$

If

$$R(\overline{x}) = \left(\limsup |x_k|^{\frac{1}{k}}\right)^{-1} > 0$$

then the series in (12) converge for all $|z| < R(\overline{x})$. So the function $F(z, \overline{x})$ is analytic for $|z| < R(\overline{x})$.

Consider an expansion in powers of z of the function $F(z, \overline{x})$:

$$F(z, \overline{x}) = \sum_{n=0}^{\infty} z^n P_n(\overline{x}).$$

One can easily show the following equalities:

$$(n+1)P_{n+1}(\overline{x}) = \sum_{k=0}^{n} (-1)^k x_{k+1} P_{n-k}(\overline{x}), \quad n \geq 0,$$

$$\frac{\partial P_n}{\partial x_l}(\overline{x}) = \begin{cases} 0 & \text{if } l > n, \\ \frac{(-1)^{l+1}}{l} P_{n-l}(\overline{x}) & \text{if } l \leq n. \end{cases}$$

4.4 Examples

1. If $\overline{x}(h) = (x, \lambda, 0, \ldots, 0, \ldots)$, then

$$F(z, \overline{x}) = \exp\left(zx - \frac{z^2}{2}\lambda\right) = \sum_{n=0}^{\infty} H_n(x, \lambda) z^n,$$

where $H_n(x, \lambda)$ are the Hermite polynomials (Brownian case). So

$$P_n(x, \lambda, 0, \ldots, 0) = H_n(x, \lambda).$$

2. If $\overline{x}(h) = (x - t, x, \ldots, x, \ldots)$, then

$$F(z, \overline{x}) = (1+z)^x e^{-tz} = \sum_{n=0}^{\infty} C_n(x, \lambda) \frac{z^n}{n!},$$

where $C_n(x, \lambda)$ are the Charlier polynomials (Poissonian case). So

$$n! P_n(x - t, x, \ldots, x) = C_n(x, \lambda).$$

4.5 Relationship Between Generalized Orthogonal Polynomials and Isonormal Lévy Process

For $h \in \mathscr{H} \cap L^{\infty}(T \times X_0, \mathscr{B}, \nu)$, let $\overline{x}(h) = (x_k(h))_{k=1}^{\infty}$ denote the sequence of the random variables such that

$$x_1(h) = L(h);$$
$$x_2(h) = L(h^2 \mathbf{1}_{X_0}) + \|h\|^2_{\mathscr{H}};$$
$$x_k(h) = L(h^k \mathbf{1}_{X_0}) + \int_{T\times X_0} h^k(t,x)\nu(dt,dx), \quad k \geq 3.$$

Lemma 4.5 *Let h and $g \in \mathscr{H} \cap L^\infty(T \times X_0, \mathscr{B}, \nu)$. Then for all $n, m \geq 0$ we have $P_n(\overline{x}(h))$ and $P_m(\overline{x}(g)) \in L^2(\Omega)$, and*

$$\mathrm{E}\left[P_n(\overline{x}(h))P_m(\overline{x}(g))\right] = \begin{cases} 0 & \text{if } n \neq m, \\ \frac{1}{n!}\left(\mathrm{E}\left[L(h)L(g)\right]\right)^n & \text{if } n = m. \end{cases}$$

4.6 The Chaos Decomposition

Lemma 4.6 *The random variables $\{e^{L(h)}, h \in \mathscr{H} \cap L^\infty(T \times X_0, \mathscr{B}, \nu)\}$ form a total subset of $L^2(\Omega, \mathscr{F}, P)$.*

For each $n \geq 1$ we will denote by $\mathscr{P}_n$ the closed linear subspace of $L^2(\Omega, \mathscr{F}, P)$ generated by the random variables $\{P_n(\overline{x}(h)), h \in \mathscr{H} \cap L^\infty(T \times X_0, \mathscr{B}, \nu)\}$. $\mathscr{P}_0$ will be the set of constants. For $n = 1$, $\mathscr{P}_1$ coincides with the set of random variables $\{L(h), h \in \mathscr{H}\}$. We will call the space $\mathscr{P}_n$ chaos of order n.

Theorem 4.7 *The space $L^2(\Omega, \mathscr{F}, P)$ can be decomposed into the infinite orthogonal sum of the subspace $\mathscr{P}_n$:*

$$L^2(\Omega, \mathscr{F}, P) = \bigoplus_{n=0}^{\infty} \mathscr{P}_n.$$

4.7 The Multiple Integral

Set $\mathscr{G}_0 = \{A \in \mathscr{G} | \pi(A) < \infty\}$. For any $m \geq 1$ we denote by $\mathscr{E}_m$ the set of all linear combinations of the following functions $f \in \mathrm{L}^2((T \times X)^m, \mathscr{G}^m, \pi^m)$

$$f(t_1, x_1, \ldots, t_m, x_m) = \mathbf{1}_{A_1 \times A_2 \times \ldots A_m}(t_1, x_1, \ldots, t_m, x_m), \tag{13}$$

where $A_1, \ldots, A_m$ are pairwise–disjoint sets in $\mathscr{G}_0$.

The fact that π is a measure without atoms implies that $\mathscr{E}_m$ is dense in $\mathrm{L}^2((T \times X)^m)$. (See, e.g. Nualart [11] pp. 8–9).

For the function of the form (13) we define the multiple integral of order m

$$I_m(f) = L(A_1) \ldots L(A_m).$$

Then, by linearity we conclude $I_m(f)$ for all functions $f \in \mathscr{E}_m$ and by continuity $I_m(f)$ for all functions $f \in \mathrm{L}^2((T \times X)^m)$.

The following properties hold:

1. I_m is linear.
2. $I_m(f) = I_m(\widetilde{f})$, where $\widetilde{f}$ denotes the symmetrization of f, which is defined by

$$\widetilde{f}(t_1, x_1, \dots, t_m, x_m) = \frac{1}{m!} \sum_{\sigma \in \mathscr{S}_m} f(t_{\sigma(1)}, x_{\sigma(1)}, \dots, t_{\sigma(m)}, x_{\sigma(m)}).$$

3.

$$\mathrm{E}\,[I_n(f) I_m(g)] = \begin{cases} 0 & if\ n \neq m, \\ m! < \widetilde{f}, \widetilde{g} >_{\mathrm{L}^2((T \times X)^m)} & if\ n = m. \end{cases}$$

4.8 Relationship Between Generalized Orthogonal Polynomials And multiple Stochastic Integrals

Proposition 4.8 *Let P_n be the nth generalized orthogonal polynomial and $\overline{x}(h) = (x_k(h))_{k=1}^{\infty}$, where $h \in \cap_{p \geq 2} L^p(T \times X_0, \mathscr{B}, \nu) \cap \mathscr{H}$ and*

$$\begin{aligned} x_1(h) &= L(h); \\ x_2(h) &= L(h^2 \mathbf{1}_{X_0}) + \|h\|^2_{\mathscr{H}}; \\ x_k(h) &= L(h^k \mathbf{1}_{X_0}) + \int_{T \times X_0} h^k(t, x) \nu(dt, dx), \quad k \geq 3. \end{aligned}$$

Then it holds that

$$n! P_n(\overline{x}(h)) = I_n(h^{\otimes n}),$$

where

$$h^{\otimes n}(t_1, x_1, \dots, t_m, x_m) = h(t_1, x_1) \times \cdots \times h(t_m, x_m).$$

4.9 Expansion into a Series of Multiple Stochastic Integrals

Corollary 4.9 *Any square integrable random variable $\xi \in L^2(\Omega, \mathscr{F}, P)$ can be expanded into a series of multiple stochastic integrals:*

$$\xi = \sum_{k=0}^{\infty} I_k(f_k). \tag{14}$$

Here $f_0 = E[\xi]$, and I_0 is the identity mapping on the constant. Furthermore, this representation is unique provided the functions $f_k \in L^2((T \times X)^k)$ are symmetric.

4.10 The Derivative Operator

Let $\mathscr{S}$ denote the class of smooth random variables such that a random variable $\xi \in \mathscr{S}$ has the form

$$\xi = f(L(h_1), \ldots, L(h_n)), \tag{15}$$

where f belongs to $C_b^\infty(\mathbb{R}^n)$, $h_1, \ldots, h_n$ are in $\mathscr{H}$, and $n \geq 1$. The set $\mathscr{S}$ is dense in $L^p(\Omega)$, for any $p \geq 1$.

Definition 4.10 The stochastic derivative of a smooth functional of the form (15) is the $\mathscr{H}$–valued random variable $D\xi = \{D_{t,x}\xi, (t,x) \in T \times X\}$ given by

$$\begin{aligned} D_{t,x}\xi = & \sum_{k=1}^{n} \frac{\partial f}{\partial y_k}(L(h_1), \ldots, L(h_n))h_k(t,x)\mathbf{1}_\Delta(x) \\ & + (f(L(h_1) + h_1(t,x), \ldots, L(h_n) + h_n(t,x)) \\ & - f(L(h_1), \ldots, L(h_n)))\, \mathbf{1}_{X_0}(x). \end{aligned} \tag{16}$$

We will consider $D\xi$ as an element of $\xi \in L^2(T \times X \times \Omega) \cong L^2(\Omega; \mathscr{H})$, namely, $D\xi$ is a random process indexed by the parameter space $T \times X$.

1. If the measure ν is zero or $h_k(t,x) = 0$, $k = 1, \ldots, n$ when $x \neq \Delta$ then $D\xi$ coincides with the Malliavin derivative (see, e.g. Nualart [11] Def. 1.2.1 p. 38).
2. If the measure μ is zero or $h_k(t,x) = 0$, $k = 1, \ldots, n$ when $x = \Delta$ then $D\xi$ coincides with the difference operator (see, e.g. Picard [13]).

4.11 Integration by Parts Formula

Theorem 4.11 *Suppose that ξ and η are smooth functionals and $h \in \mathscr{H}$. Then*

1.

$$\mathrm{E}[\xi L(h)] = \mathrm{E}[\langle D\xi; h\rangle_{\mathscr{H}}].$$

2.

$$\mathrm{E}[\xi\eta L(h)] = \mathrm{E}[\eta\,\langle D\xi; h\rangle_{\mathscr{H}}] + \mathrm{E}[\xi\,\langle D\eta; h\rangle >_{\mathscr{H}}] + \mathrm{E}[\big(D\eta; h\mathbf{1}_{X_0} D\xi\big\rangle_{\mathscr{H}}].$$

As a consequence of the above theorem we obtain the following result:

- The expression of the derivative $D\xi$ given in (16) does not depend on the particular representation of ξ in (15).
- The operator D is closable as an operator from $L^2(\Omega)$ to $L^2(\Omega; \mathscr{H})$.

We will denote the closure of D again by D and its domain in $L^2(\Omega)$ by $\mathbb{D}^{1,2}$.

4.12 The Chain Rule

Proposition 4.12 (See Yablonski [16], Proposition 4.8) *Suppose $F = (F_1, F_2, \ldots, F_n)$ is a random vector whose components belong to the space $\mathbb{D}^{1,2}$. Let $\phi \in \mathscr{C}^1(\mathbb{R}^n)$ be a function with bounded partial derivatives such that $\phi(F) \in \mathrm{L}^2(\Omega)$. Then $\phi(F) \in \mathbb{D}^{1,2}$ and*

$$D_{t,x}\phi(F) = \begin{cases} \sum_{i=1}^{n} \frac{\partial \phi}{\partial x_i}(F) D_{t,\Delta} F_i; & x = \Delta \\ \phi(F_1 + D_{t,x}F_1, \ldots, F_n + D_{t,x}F_n) - \phi(F_1, \ldots, F_n); & x \neq \Delta \end{cases}$$

4.13 The Action of the Operator D via the Chaos Decomposition

Lemma 4.13 *It holds that $P_n(\overline{x}(h)) \in \mathbb{D}^{1,2}$ for all $h \in \mathscr{H} \cap L^\infty(T \times X_0, \mathscr{B}, \nu)$, $n = 1, 2, \ldots$ and*

$$D_{t,x} P_n(\overline{x}(h)) = P_{n-1}(\overline{x}(h)) h(t, x).$$

Proposition 4.14 *Let $\xi \in L^2(\Omega, \mathscr{F}, P)$ with an expansion $\xi = \sum_{k=0}^{\infty} I_k(f_k)$ where $f_k \in L^2((T \times X)^k)$ are symmetric for all k. Then $\xi \in \mathbb{D}^{1,2}$ if and only if*

$$\sum_{k=0}^{\infty} k k! \| f_k \|^2_{L^2((T\times X)^k)} < \infty,$$

and in this case we have

$$D_{t,x}\xi = \sum_{k=0}^{\infty} k I_{k-1}(f_k(\cdot, t, x))$$

and

$$\mathrm{E}\left[\int_{T\times X}(D_{t,x}\xi)^2\pi(dt,dx)\right]$$

coincides with the sum of the series (14).

4.14 The Skorohod Integral

We recall that the derivative operator D is a closed and unbounded operator defined on the dense subset $\mathbb{D}^{1,2}$ of $\mathrm{L}^2(\Omega)$ with values in $\mathrm{L}^2(\Omega;\mathscr{H})$.

Definition 4.15 We denote by δ the adjoint of the operator D and we call it the Skorohod integral.
The operator δ is a closed and unbounded operator on $\mathrm{L}^2(\Omega;\mathscr{H})$ with values in $\mathrm{L}^2(\Omega)$ defined on $Dom(\delta)$, where $Dom(\delta)$ is the set of processes $u\in\mathrm{L}^2(\Omega;\mathscr{H})$ such that

$$\left|\mathrm{E}\left[\int_{T\times X}D_{t,z}Fu(t,z)\pi(dt,dz)\right]\right|\leq c\,\|F\|_{\mathrm{L}^2(\Omega)}$$

for all $F\in\mathbb{D}^{1,2}$, where c is some constant depending on u.

If $u\in Dom(\delta)$, then $\delta(u)$ is the element of $\mathrm{L}^2(\Omega)$ such that

$$\mathrm{E}\left[F\delta(u)\right]=\mathrm{E}\left[\int_{T\times X}D_{t,z}Fu(t,z)\pi(dt,dz)\right] \tag{17}$$

for any $F\in\mathbb{D}^{1,2}$.

4.15 The Behavior of δ in Terms of the Chaos Expansion

Proposition 4.16 *Let $u\in\mathrm{L}^2(\Omega;\mathscr{H})$ with the expansion*

$$u(t,z)=\sum_{k=0}^{\infty}I_k(f_k(\cdot,t,z)). \tag{18}$$

Then $u\in Dom(\delta)$ if and only if the series

$$\delta(u)=\sum_{k=0}^{\infty}I_{k+1}(\widetilde{f_k}) \tag{19}$$

converges in $\mathrm{L}^2(\Omega)$.

It follows that $Dom(\delta)$ is the subspace of $L^2(\Omega)$ formed by the processes that satisfy the following condition:

$$\sum_{k=1}^{\infty}(k+1)!\|\widetilde{f}_k\|^2_{L^2(T\times X)^{k+1}} < \infty. \tag{20}$$

Note that the Skorohod integral is a linear operator and has a zero mean, e.g. $\mathbb{E}[\delta(u)] = 0$ if $u \in Dom(\delta)$. The following statements prove some properties of δ.

Proposition 4.17 *Suppose that u is a Skorohod integrable process. Let $F \in \mathbb{D}^{1,2}$ be such that* $\mathrm{E}\left[\int_{T\times X}\left(F^2 + (D_{t,z}F)^2 1_{X_0}\right)u(t,z)^2\pi(dt,dz)\right] < \infty$. *Then it holds that*

$$\delta\left(\left(F + (D_{t,z}F)1_{X_0}\right)u\right) = F\delta(u) - \int_{T\times X}(D_{t,z}F)u(t,z)\pi(dt,dz), \tag{21}$$

provided that one of the two sides of the equality (21) exists.

4.16 Commutativity Relationship Between the Derivative and Divergence Operators

Let $\mathbb{L}^{1,2}$ denote the class of processes $u \in L^2(T \times X \times \Omega)$ such that $u(t,x) \in \mathbb{D}^{1,2}$ for almost all (t,x), and there exists a measurable version of the multi–process $D_{t,x}u(s,y)$ satisfying

$$\mathrm{E}\left[\int_{T\times X}\int_{T\times X}(D_{t,x}u(s,y))^2\pi(dt,dx)\pi(dsdy)\right] < \infty.$$

Proposition 4.18 *Suppose that $u \in \mathbb{L}^{1,2}$ and for almost all $(t,z) \in T \times X$, the two–parameter process $\left(D_{t,z}u(s,y)\right)_{(s,y)\in T\times X}$ is Skorohod integrable, and there exists a version of the process $\left(\delta(D_{t,z}u(\cdot,\cdot))\right)_{(t,z)\in T\times X}$ which belongs to $L^2(T\times X\times\Omega)$. Then $\delta(u) \in \mathbb{D}^{1,2}$, and we have*

$$D_{t,z}\delta(u) = u(t,z) + \delta(D_{t,z}u(\cdot,\cdot)). \tag{22}$$

4.17 The Itô Stochastic Integral as a Particular Case of the Skorohod Integral

Let $W = \{W_t, 0 \le t \le T\}$ is a be an d-dimensional standard Brownian motion, $\widetilde{N}$ a compensated Poisson random measure on $[0,T] \times \mathbb{R}^d_0$ with (*time-inhomogeneous*) intensity measure $\nu(dt,dx) = \beta_t(dx)dt$, where $(\beta_t)_{t\in[0,T]}$ is a family of Lévy measures verifying $\int_0^T \left(\int_{\mathbb{R}^d}(\|z\|^2 \wedge 1)\beta_t(dz)\right) dt < \infty$. Here $\mathbb{R}_0 := \mathbb{R} \setminus \{0\}$ and for each $t \in [0, T]$, $\mathscr{F}_t$ is the σ–algebra generated by the random variables

$$\{W_s^j, \widetilde{N}((0, s] \times A); 0 \leq s \leq t, j = 1, \ldots, d, \ A \in \mathscr{B}(\mathbb{R}_0^d), \ \sup_{0 \leq s \leq t} \beta_s(A) < \infty\}$$

and the null sets of $\mathscr{F}$.
We denote by L_p^2 the subset of $L^2(\Omega; \mathscr{H})$ formed by $(\mathscr{F}_t)$–predictable processes.

Proposition 4.19 *$L_p^2 \subset Dom(\delta)$, and the restriction of the operator δ to the space coincides with the usual stochastic integral, that is*

$$\delta(u) = \sum_{j=1}^{d} \int_0^T u^j(t, 0) dW_t^j + \int_0^T \int_{\mathbb{R}_0^d} u(t, z)\widetilde{N}(dt, dz). \tag{23}$$

References

1. Bally, V., Bavouzet, M.-P., Massaoud, M.: Integration by parts formula for locally smooth laws and applications to sensitivity computations. Ann. Appl. Probab. **17**(1), 33–66 (2007)
2. Benth, F.E., Di Nunno, G., Khedher, A.: Robustness of option prices and their deltas in markets modelled by jump-diffusions. Commun. Stochast. Anal. **5**(2), 285–307 (2011)
3. Cont, R., Tankov, P.: Financial Modelling with Jump Processes. Chapman & Hall/CRC, Boca Raton (2004)
4. Cox, J.C., Ingersoll, J.E., Ross, S.A.: A theory of the term structure of interest rates. Econometrica **53**, 385–407 (1985)
5. Davis, M.A., Johansson, M.P.: Malliavin Monte Carlo Greeks for jump diffusions. Stochast. Process. Appl. **116**(1), 101–129 (2006)
6. Di Nunno, G., Øksendal, B., Proske, F.: Malliavin Calculus for Lévy Processes with Applications to Finance, Springer (2008)
7. El-Khatib, Y., Privault, N.: Computations of greeks in a market with jumps via the malliavin calculus. Finance Stochast. **8**(2), 161–179 (2004)
8. El-Khatib, Y., Hatemi, J.A.: On the calculation of price sensitivities with jump-diffusion structure. J. Stat. Appl. Probab. **3**(1), 171–182 (2012)
9. Fournié, E., Lasry, J.-M., Lebuchoux, J., Lions, P.-L., Touzi, N.: Applications of Malliavin calculus to Monte Carlo methods in finance. Finance Stochast. **3**(4), 391–412 (1999)
10. León, J.A., Solé, J.L., Utzet, F., Vives, J.: On Lévy processes, Malliavin calculus and market models with jumps. Finance Stochast. **6**, 197–225 (2002)
11. Nualart, D.: The Malliavin Calculus and Related Topics, 2nd edn. Springer, Berlin Heidelberg (2006)
12. Petrou, E.: Malliavin calculus in Lévy spaces and applications to finance. Electron. J. Probab. **13**(27), 852–879 (2008)
13. Picard, J.: On the existence of smooth densities for jump processes. Probab. Theory Relat. Fields **105**(4), 481–511 (1996)
14. Protter, P.: Stochastic Integration and Differential Equations. Stochastic Modeling and Applied Probability, vol. 21. Springer, Berlin (2005)
15. Sato, K.-I.: Lévy Processes and Infinitely Divisible Distributions. Cambridge University Studies in Advanced Mathematics, vol. 68. Cambridge University Press, Cambridge (1999)
16. Yablonski, A.: The calculus of variations for processes with independent increments. Rocky Mt J. Math. **38**, 669–701 (2008)

Variance-GGC Asset Price Models and Their Sensitivity Analysis

Nicolas Privault and Dichuan Yang

Abstract This paper reviews the variance-gamma asset price model as well as its symmetric and non-symmetric extensions based on generalized gamma convolutions (GGC). In particular we compute the basic characteristics and decomposition of the variance-GGC model, and we consider its sensitivity analysis based on the approach of Kawai and Kohatsu-Higa in Appl Math Finance 17(4):301–321, 2010 [8].

Keywords Variance-gamma model · Variance-GGC model · Sensitivity analysis

Mathematics Subject Classification 60E07 · 60G51 · 60J65 · 60G52 · 62P05 · 91B28

1 Introduction

Lévy processes play an important role in the modeling of risky asset prices with jumps. In addition to the Black-Scholes model based on geometric Brownian motion, pure jump and jump-diffusion processes have been used by Cox and Ross [5] and Merton [13] for the modeling of asset prices. More recently, Brownian motions time-changed by non-decreasing Lévy processes (i.e. subordinators) have become popular, in particular the Normal Inverse Gaussian (NIG) model [1], the variance-gamma (VG) model [11, 12], and the CGMY/KoBol models [3, 4].

N. Privault (✉)
Division of Mathematical Sciences, School of Physical and Mathematical Sciences, Nanyang Technological University, 21 Nanyang Link, 637371 Nanyang, Singapore
e-mail: nprivault@ntu.edu.sg

D. Yang
Department of Mathematics, City University of Hong Kong, Tat Chee Avenue, Kowloon Tong, Hong Kong
e-mail: dyang5@student.cityu.edu.hk

M. Eddahbi et al. (eds.), *Statistical Methods and Applications in Insurance and Finance*, Springer Proceedings in Mathematics & Statistics 158,
DOI 10.1007/978-3-319-30417-5_3

The normal inverse Gaussian (NIG) process [1] can constructed as a Brownian motion time-changed by a Lévy process with the inverse Gaussian distribution, whose marginal at time t is identical in law to the first hitting time of the positive level t by a drifted Brownian motion.

The variance-gamma process [11, 12] is built on the time change of a Brownian motion by a gamma process, and has been successful in modeling asset prices with jumps and in addressing the issue of slowly decreasing probability tails found in real market data.

The CGMY/KoBol models [3, 4] are extensions of the variance-gamma model by a more flexible choice of Lévy measures. However, this extension loses some nice properties of variance-gamma model, for example variance-gamma processes can be decomposed into the difference of two gamma processes, whereas this property does not hold in general in the CGMY/KoBol models.

In [6] the variance-gamma model has been extended into a symmetric variance-GGC model, based on generalized gamma convolutions (GGCs), see [2] for details and a driftless Brownian motion. In this paper we review this model and propose an extension to non-symmetric case using a drifted Brownian motion.

GGC random variables can be constructed by limits in distribution of sums of independent gamma random variables with varying shape parameters. As a result, the variance-GGC model allows for more flexibility than standard variance-gamma models, while retaining some of their properties. The skewness and kurtosis of variance-GGC processes can be computed in closed form, including the relations between skewness and kurtosis of the GGC process and of the corresponding variance-GGC process. In addition, variance-GGC processes can be represented as the difference of two GGC processes.

On the other hand, the sensitivity analysis of stochastic models is an important topic in financial engineering applications. The sensitivity analysis of time-changed Brownian motion processes has been developed and the Greek formulas have been obtained by following the approach in [8]. In addition, the sentivity analysis of the variance-gamma, stable and tempered stable processes has been performed in [9] and [10] respectively. As an extension of the variance-gamma process, we study the corresponding sensitivity analysis of the variance-GGC model along the lines of [9].

In the remaining of this section we review some facts on generalized gamma convolutions, (GGCs) including their variance, skewness and kurtosis. We also discuss an asset price model based on GGCs and its sensitivity analysis.

Wiener-gamma integrals

Consider a gamma process $(\gamma_t)_{t\in\mathbb{R}_+}$, i.e. $(\gamma_t)_{t\in\mathbb{R}_+}$ is a process with independent and stationary increments such that γ_t at time $t > 0$ has a gamma distribution with shape parameter t and probability density function $e^{-x}x^{t-1}/\Gamma(t)$, $x > 0$. We denote by

$$\int_0^\infty g(t)d\gamma_t, \tag{1}$$

the Wiener-gamma stochastic integral of a deterministic function

$$g : \mathbb{R}_+ \longrightarrow \mathbb{R}_+$$

with respect to the standard gamma process $(\gamma_t)_{t\in\mathbb{R}_+}$, provided g satisfies the condition

$$\int_0^\infty \log(1+g(t))dt < \infty, \qquad (2)$$

which ensures the finiteness of Eq. (1), cf. Sect. 1.2, page 350 of [7] for details. In particular, there is a one-to-one correspondence between GGC random variables and Wiener-gamma integrals, Proposition 1.1, page 352 of [7].

Generalized gamma convolutions

A random variable Z is a generalized gamma convolution if its Laplace transform admits the representation

$$\mathbb{E}[e^{-uZ}] = \exp\left(-t\int_0^\infty \log\left(1+\frac{u}{s}\right)\mu(ds)\right), \qquad u \geq 0$$

where $\mu(ds)$ is called the Thorin measure and should satisfy the conditions

$$\int_{(0,1]} |\log s|\mu(ds) < \infty \quad \text{and} \quad \int_{(1,\infty)} s^{-1}\mu(ds) < \infty.$$

Generalized gamma convolutions (GGC) can be defined as the limits of independent sums of gamma random variables with various shape parameters, cf. [2] for details.

In particular, the density of the Lévy measure of a GGC random variable is a completely monotone function. From the Laplace transform of Z we find

$$\mathbb{E}[Z] = \int_0^\infty t^{-1}\mu(dt),$$

and the first central moments of Z can be computed as

$$\begin{cases} \mathbb{E}[(Z-\mathbb{E}[Z])^2] = \displaystyle\int_0^\infty t^{-2}\mu(dt), \\ \mathbb{E}[(Z-\mathbb{E}[Z])^3] = 2\displaystyle\int_0^\infty t^{-3}\mu(dt), \\ \mathbb{E}[(Z-\mathbb{E}[Z])^4] = 3\,(\mathrm{Var}\,[Z])^2 + 6\displaystyle\int_0^\infty t^{-4}\mu(dt). \end{cases} \qquad (3)$$

As a consequence we can compute the

$$\text{Skewness}[Z] = \frac{\mathbb{E}[(Z - \mathbb{E}[Z])^3]}{(\text{Var}[Z])^{3/2}} = \frac{2\int_0^\infty t^{-3}\mu(dt)}{(\text{Var}[Z])^{3/2}},$$

and

$$\text{Kurtosis}[Z] = \frac{\mathbb{E}[(Z - \mathbb{E}[Z])^4]}{(\text{Var}[Z])^2} = 3 + 6\frac{\int_0^\infty t^{-4}\mu(dt)}{(\text{Var}[Z])^2}$$

of Z. We refer the reader to Proposition 1.1 of [7] for the relation between the integrand in a Wiener-gamma representation and the cumulative distribution function of the associated generalized gamma convolution.

Market model and sensitivity analysis

As an extension of the model of [9] to GGC random variables we consider an asset price process S_T defined by the exponent

$$S_T = S_0 \exp\left(\theta \int_0^\infty g(s)d\gamma_s + \tau\sqrt{T}\Theta + Z_T + c(\theta,\tau)T\right),$$

of a variance-GGC process, i.e. $\int_0^\infty g(s)d\gamma_s$ is a GGC random variable represented as a Wiener-gamma integral, Θ is an independent Gaussian random variable, $(Z_t)_{t\in\mathbb{R}_+}$ is another GGC-Lévy process, and $\theta \in \mathbb{R}$, $\tau \geq 0$, $T > 0$.

In Sect. 3 the sensitivity $\frac{\partial}{\partial S_0}\mathbb{E}[\Phi(S_T)]$ of an option with payoff Φ with respect to the initial value S_0 in a variance-GGC model is shown to satisfy

$$\frac{\partial}{\partial S_0}\mathbb{E}[\Phi(S_T)] = \frac{1}{S_0}\mathbb{E}[\Phi(S_T)L_T],$$

where

$$L_T := \frac{2\theta\int_0^\infty g(s)f^2(s)d\gamma_s}{(\theta\int_0^\infty g(s)f(s)d\gamma_s + \tau\sqrt{T}\eta)^2} + \frac{\int_0^\infty f(s)d\gamma_s - T\int_0^\infty f(s)ds + \eta\Theta}{\theta\int_0^\infty g(s)f(s)d\gamma_s + \tau\sqrt{T}\eta}$$

for any positive function $f : \mathbb{R}_+ \to (0, a)$ and $\eta > 0$. In Theorem 1 we will compute this sensitivity as well as orther Greeks based on the model parameters θ and τ.

The remaining of this paper is organized as follows. In Sect. 2 we introduce a model for Brownian motion time-changed by a GGC subordinator. The variance, skewness and kurtosis of variance-GGC processes are calculated in relation with the corresponding parameters of GGC processes, and several example of variance-GGC models are considered. A Girsanov transform of GGC processes is also stated. The sensitivity analysis with respect to S_0, θ and τ is conducted in Sect. 3.

2 Variance-GGC Processes

Given $(W_t)_{t\in\mathbb{R}_+}$ a standard Brownian motion and $\theta \in \mathbb{R}$, $\sigma > 0$, consider the drifted Brownian motion

$$B_t^{\theta,\sigma} := \theta t + \sigma W_t, \qquad t \in \mathbb{R}_+.$$

Next, consider a generalized gamma convolution (GGC) Lévy process $(G_t)_{t\in\mathbb{R}_+}$ such that G_1 is a GGC random variable with Thorin measure $\mu(ds)$ on $\mathbb{R}_+$. We define the variance-GGC process $(Y_t^{\sigma,\theta})_{t\in\mathbb{R}_+}$ as the time-changed Brownian motion

$$Y_t^{\sigma,\theta} := B_{G_t}^{\theta,\sigma}, \qquad t \in \mathbb{R}_+.$$

The probability density function of $Y_t^{\sigma,\theta}$ is given by

$$f_{Y_t^{\sigma,\theta}}(x) = \frac{1}{\sigma\sqrt{2\pi}} \int_0^\infty \exp\left(-\frac{|x-\theta y|^2}{2\sigma^2 y}\right) h_t(y) \frac{dy}{\sqrt{y}}, \qquad x \in \mathbb{R},$$

where $h_t(y)$ is the probability density function of G_t, cf. Relation (6) in [11].

The Laplace transform of $Y_t^{\sigma,\theta}$ is

$$\begin{aligned}
\mathbb{E}\left[\exp\left(-uY_t^{\sigma,\theta}\right)\right] &= \int_0^\infty e^{-uy} f_{Y_t}(y) dy \\
&= \Psi_{G_t}\left(\theta u - \frac{\sigma^2}{2}u^2\right) \\
&= \exp\left(-t\int_0^\infty \log\left(1 + \frac{\theta u - \sigma^2 u^2/2}{s}\right)\mu(ds)\right), \qquad (4)
\end{aligned}$$

where Ψ_{G_t} is the Laplace transform of G_t.

This construction extends the symmetric variance-GGC model constructed in Sect. 4.4, pages 124–126 of [6]. In particular, the next proposition extends to variance-GGC processes Relation (8) in [11, 12], which decomposes the variance-gamma process into the difference of two gamma processes. Here, we are writing Y_t as the difference of two independent GGC processes, i.e. Y_t becomes an Extended Generalized Gamma Convolution (EGGC) in the sense of Chap. 7 of [2], cf. also Sect. 3 of [14].

Proposition 1 *The time-changed process Y_t can be decomposed as*

$$Y_t = U_t - W_t,$$

where U_t and W_t are two independent GGC processes with Thorin measures μ_A and μ_B which are the image measures of $\mu(dt)$ on $\mathbb{R}_+$ respectively, by the mappings

$$s \longmapsto B(s) := \frac{\theta}{\sigma^2} + \frac{1}{\sigma}\sqrt{\frac{\theta^2}{\sigma^2} + 2s}, \qquad s \in \mathbb{R}_+,$$

and

$$s \longmapsto A(s) = -\frac{\theta}{\sigma^2} + \frac{1}{\sigma}\sqrt{\frac{\theta^2}{\sigma^2} + 2s}, \qquad s \in \mathbb{R}_+.$$

Proof From (4), the Laplace tranform of Y_t can be decomposed as

$$\begin{aligned}
&\mathbb{E}\left[\exp\left(-uY_t^{\sigma,\theta}\right)\right] = \exp\left(-t\int_0^\infty \log\left(1 - \frac{u}{B(s)}\right)\left(1 + \frac{u}{A(s)}\right)\mu(ds)\right) \\
&= \exp\left(-t\int_0^\infty \log\left(1 + \frac{u}{A(s)}\right)\mu(ds) - t\int_0^\infty \log\left(1 - \frac{u}{B(s)}\right)\mu(ds)\right) \\
&= \exp\left(-t\int_0^\infty \log\left(1 + \frac{u}{s}\right)\mu_A(ds) - t\int_0^\infty \log\left(1 - \frac{u}{s}\right)\mu_B(ds)\right) \\
&= \mathbb{E}[e^{-uU_t}]\mathbb{E}[e^{uW_t}].
\end{aligned}$$

□

The Laplace tranform of Y_t can also be decomposed as

$$\begin{aligned}
\mathbb{E}\left[\exp\left(-uY_t^{\sigma,\theta}\right)\right] &= \exp\left(-t\int_0^\infty \log\left(1 + \frac{u}{s}\right)\mu_A(ds) - t\int_0^\infty \log\left(1 - \frac{u}{s}\right)\mu_B(ds)\right) \\
&= \exp\left(-t\int_{-\infty}^0 \log\left(1 + \frac{u}{s}\right)\mu_{-B}(ds) - t\int_0^\infty \log\left(1 + \frac{u}{s}\right)\mu_A(ds)\right), \qquad (5)
\end{aligned}$$

where μ_{-B} is the image measure of μ_B by $s \mapsto -s$, and in particular, Y_t is an extended GGC (EGGC) with Thorin measure $\mu_A + \mu_{-B}$ in the sense of Chap. 7 of [2].

In the next proposition we compute the variance, skewness and kurtosis of variance-GGC processes.

Proposition 2 *We have*

(*i*) $\mathrm{Var}[Y_1] = \theta^2 \mathrm{Var}[G_1] + \sigma^2 \mathbb{E}[G_1]$.

(*ii*) $$\begin{aligned}
\mathrm{Skewness}[Y_1] &= -\frac{\mathbb{E}[(G_1 - \mathbb{E}[G_1])^3] + 2(\sigma/\theta)^2 \mathrm{Var}[G_1]}{2(\mathrm{Var}[G_1] + (\sigma/\theta)^2 \mathbb{E}[G_1])^{3/2}} \\
&= -\frac{\theta^3}{2}\mathrm{Skewness}[G_1]\frac{(\mathrm{Var}[G_1])^{3/2}}{(\mathrm{Var}[Y_1])^{3/2}} - \frac{\theta\sigma^2 \mathrm{Var}[G_1]}{(\mathrm{Var}[Y_1])^{3/2}}. \qquad (6)
\end{aligned}$$

$$(iii)\ \ \text{Kurtosis}[Y_1] = 3 + 3\theta^4 \frac{\mathbb{E}[(G_1 - \mathbb{E}[G_1])^4] - 3(\text{Var}[G_1])^2}{8(\theta^2 \text{Var}[G_1] + \sigma^2 \mathbb{E}[G_1])^2}$$
$$+ 3\frac{3\theta^2\sigma^2\mathbb{E}[(G_1 - \mathbb{E}[G_1])^3]/4 + \sigma^4 \text{Var}[G_1]}{(\theta^2 \text{Var}[G_1] + \sigma^2 \mathbb{E}[G_1])^2}$$
$$= 3 + \theta^4 \frac{(\text{Kurtosis}[G_1] - 3)(\text{Var}[G_1])^2}{16(\text{Var}[Y_1])^2}$$
$$+ 9\sigma^2\theta^2 \frac{\text{Skewness}[G_1](\text{Var}[G_1])^{3/2}}{4(\text{Var}[Y_1])^2} + 3\frac{\sigma^4 \text{Var}[G_1]}{(\text{Var}[Y_1])^2}. \tag{7}$$

Proof Using the Thorin measure $\mu_A + \mu_{-B}$ of Y_t and (3) we have

$$\begin{aligned}
\text{Var}[Y_1] &= \int_0^\infty t^{-2}\mu_A(dt) + \int_{-\infty}^0 t^{-2}\mu_{-B}(dt) \\
&= \int_0^\infty \frac{1}{A^2(t)}\mu(dt) + \int_0^\infty \frac{1}{B^2(t)}\mu(dt) \\
&= \int_0^\infty \frac{\theta^2 + t\sigma^2}{t^2}\mu(dt) \\
&= \theta^2 \text{Var}[G_1] + \sigma^2 \mathbb{E}[G_1],
\end{aligned}$$

and

$$\begin{aligned}
\mathbb{E}[(Y_1 - \mathbb{E}[Y_1])^3] &= 2\int_0^\infty t^{-3}\mu_A(dt) + 2\int_{-\infty}^0 t^{-3}\mu_{-B}(dt) \\
&= \frac{1}{2}\int_0^\infty \frac{\theta^3 + \theta\sigma^2\left(\theta^2/\sigma^2 + 2t\right)}{t^3}\mu(dt) \\
&= \frac{\theta^3}{2}\mathbb{E}[(G_1 - \mathbb{E}[G_1])^3] + \theta\sigma^2 \text{Var}[G_1],
\end{aligned}$$

and

$$\begin{aligned}
\mathbb{E}[(Y_1 - \mathbb{E}[Y_1])^4] &= 6\int_{-\infty}^0 t^{-4}\mu_A(dt) + 6\int_0^\infty t^{-4}\mu_{-B}(dt) \\
&\quad + 3\left(\int_{-\infty}^0 t^{-2}\mu^-(dt) + \int_0^\infty t^{-2}\mu^+(dt)\right)^2 \\
&= \frac{3}{4}\int_0^\infty \frac{\theta^4 + (\theta\sigma)^2(\sqrt{4\theta^2/\sigma^2 + 8t}/2)^2 + \sigma^4(\sqrt{4\theta^2/\sigma^2 + 8t})^4/2}{t^4}\mu(dt) \\
&\quad + 3\left(\int_0^\infty \frac{\theta^2 + t\sigma^2}{t^2}\mu(dt)\right)^2 \\
&= \frac{3}{4}\int_0^\infty \frac{3\theta^4 + 6\sigma^2\theta^2 t + 4\sigma^4 t^2}{t^4}\mu(dt) + 3\left(\int_0^\infty \frac{\theta^2 + t\sigma^2}{t^2}\mu(dt)\right)^2
\end{aligned}$$

$$= \frac{3}{8}\theta^4(\mathbb{E}[(G_1 - \mathbb{E}[G_1])^4] - 3(\mathrm{Var}[G_1])^2)$$
$$+\frac{9}{4}\theta^2\sigma^2\mathbb{E}[(G_1 - \mathbb{E}[G_1])^3] + 3\sigma^4\mathrm{Var}[G_1] + 3(\theta^2\mathrm{Var}[G_1] + \sigma^2\mathbb{E}[G_1])^2,$$

and this yields (6) and (7). □

Girsanov theorem

Consider the probability measure Q_λ defined by the Radon-Nikodym density

$$\frac{dQ_\lambda}{dP} := \frac{e^{\lambda Y_T}}{\mathbb{E}[e^{\lambda Y_T}]} = (1-\lambda)^{aT} e^{\lambda Y_T} = e^{\lambda Y_T + aT\log(1-\lambda)}, \qquad \lambda < 1, \tag{8}$$

cf. e.g. Lemma 2.1 of [9], where Y_T is a gamma random variable with shape and scale parameters $(aT, 1)$ under P. Then, under Q_λ, the random variable Y_t has a gamma distribution with parameter $(aT, 1/(1-\lambda))$, i.e. the distribution of $Y_t/(1-\lambda)$ under P.

In the next proposition we extend this Girsanov transformation to GGC random variables.

Proposition 3 *Consider the probability measure P_f defined by its Radon-Nikodym derivative*

$$\frac{dP_f}{dP} = \frac{e^{\int_0^\infty f(s)d\gamma_s}}{\mathbb{E}[e^{\int_0^\infty f(s)d\gamma_s}]} = e^{\int_0^\infty f(s)d\gamma_s + \int_0^\infty \log(1-f(s))ds},$$

where $f : \mathbb{R}_+ \to (0,1)$ satisfies

$$\int_0^\infty \log\left(\frac{1+f(t)}{1-f(t)}\right) dt < \infty. \tag{9}$$

Assume that $g : \mathbb{R}_+ \to \mathbb{R}_+$ satisfies (2), and

$$\int_0^\infty \log(1 + ug(s) - f(s))\, ds > -\infty, \qquad u > 0.$$

Then, under P_f, the law of $\int_0^\infty g(s)d\gamma_s$ is the GGC distribution of the Wiener-gamma integral

$$\int_0^\infty \frac{g(s)}{1-f(s)} d\gamma_s$$

under P.

Proof For all $u > 0$, we have

$$\begin{aligned}
&\mathbb{E}_{P_f}\left[\exp\left(-u\int_0^\infty g(s)d\gamma_s\right)\right]\\
&= \mathbb{E}\left[\exp\left(-u\int_0^\infty g(s)d\gamma_s + \int_0^\infty f(s)d\gamma_s + \int_0^\infty \log(1-f(s))\,ds\right)\right]\\
&= \mathbb{E}\left[\exp\left(\int_0^\infty f(s) - ug(s)d\gamma_s\right)\right]\exp\left(\int_0^\infty \log(1-f(s))\,ds\right)\\
&= \exp\left(-\int_0^\infty \log(1+ug(s)-f(s))\,ds\right)\exp\left(\int_0^\infty \log(1-f(s))\,ds\right)\\
&= \exp\left(-\int_0^\infty \log\left(1+\frac{ug(s)}{1-f(s)}\right)ds\right)\\
&= \mathbb{E}\left[\exp\left(-u\int_0^\infty \frac{g(s)}{1-f(s)}d\gamma_s\right)\right].
\end{aligned}$$

□

Note that (8) is recovered by taking $g(s) = \mathbf{1}_{[0,aT]}(s)$ and $f(s) = \lambda\mathbf{1}_{[0,aT]}(s)$ for $\lambda \in (0,1)$, i.e. $G_T = \int_0^\infty g(s)d\gamma_s$ is a gamma random variable with shape parameter aT and we have

$$\mathbb{E}_{P_f}[e^{-uG_T}] = \left(1+\frac{u}{1-\lambda}\right)^{-aT} = \mathbb{E}\left[\exp\left(-\frac{u}{1-\lambda}G_T\right)\right],$$

$u > 0$, $\lambda < 1$. Next we consider several examples and particular cases.

Gamma case

In case the Thorin measure μ is given by

$$\mu(dt) = \gamma\delta_c(dt),$$

where δ_c is the Dirac measure at $c > 0$ we find the variance-gamma model of [12]. Here, G_t, $t > 0$, has the gamma probability density

$$\phi_t(x) = c^{\gamma t}\frac{x^{\gamma t-1}e^{-cx}}{\Gamma(\gamma t)}, \qquad x \in \mathbb{R}_+,$$

with mean and variance $\gamma t/c$ and $\gamma t/c^2$, and G_t becomes a gamma random variable with parameters $(\gamma t, c)$. In this case, the decomposition in Proposition 1 reads

$$\Psi_{Y_t}(u) = \left(1-\frac{\sigma^2u^2}{2c}\right)^{-t\gamma} = \left(1-\frac{\sigma u}{\sqrt{2c}}\right)^{-t\gamma}\left(1+\frac{\sigma u}{\sqrt{2c}}\right)^{-t\gamma},$$

and we have

$$\mu_A(dt) = \mu_B(dt) = \gamma\delta_{\sqrt{2c}/\sigma}(dt),$$

thus $(U_t)_{t\in\mathbb{R}_+}$, $(W_t)_{t\in\mathbb{R}_+}$ become independent gamma processes with parameter $(\gamma t, \sqrt{2c}/\sigma)$. The mean and variance of U_1 are

$$\mathbb{E}[U_1] = \int_0^\infty t^{-1}\mu_A(dt) = \frac{\sigma\gamma}{\sqrt{2c}}$$

and

$$\mathrm{Var}[U_1] = \mathbb{E}[(U_1 - \mathbb{E}[U_1])^2] = \int_0^\infty t^{-2}\mu_A(dt) = \frac{\gamma\sigma^2}{2c}.$$

Symmetric case

When $\theta = 0$ we recover the symmetric variance-GGC process

$$Y_t := B^\sigma(G_t), \qquad t \in \mathbb{R}_+,$$

defined in Sect. 4.4, page 124–126 of [6], i.e. the time-changed Brownian motion is a symmetric variance-GGC process. Here, Y_t is a centered Gaussian random variable with variance $\sigma^2 G_t$ given G_t, where B_t^σ is a standard Brownian motion with variance σ^2.

The Laplace transform of Y_t in Proposition 1 shows that Y_t decomposes into two independent processes with same GGC increments since μ_A and μ_B are the same image measures of $\mu(dt)$ on $\mathbb{R}_+$, by $s \mapsto \sqrt{2s}/\sigma$.

Variance-stable processes

Let $(G_t)_{t\in\mathbb{R}_+}$ be a Lévy stable process with index parameter $\alpha \in (0, 1)$ and moment generating function $h(s) = e^{-s^\alpha}$. In this section we consider a non-symmetric extension of the symmetric variance stable process considered in Sect. 4.5, pages 126–127 of [6]. The Thorin measure of the stable distribution is given by

$$\mu(dt) = \varphi(t)dt = \frac{\alpha}{\pi}\sin(\alpha\pi)t^{\alpha-1}dt,$$

cf. page 35 of [2]. By Proposition 1, Y_t can be decomposed as

$$Y_t = U_t - W_t,$$

where U_t and W_t are processes with independent stable increments and Thorin measures

$$\mu_A(dt) = \varphi_A(t)dt = \frac{\alpha}{\pi}\sin(\alpha\pi)(\sigma^2 t + \theta)\left(\frac{1}{2}(\sigma t - \theta/\sigma)^2 - \frac{\theta^2}{2\sigma^2}\right)^{\alpha-1} dt,$$

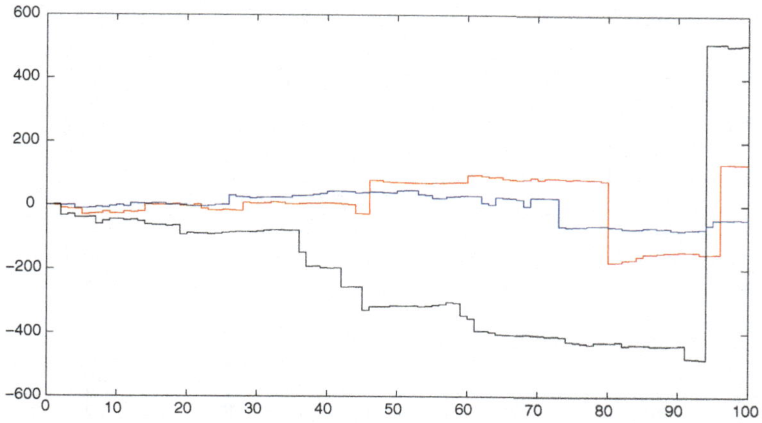

Fig. 1 Sample paths of variance-stable process with $\alpha = 0.99$

and

$$\mu_B(dt) = \varphi_B(t)dt = \frac{\alpha}{\pi}\sin(\alpha\pi)(\sigma^2 t - \theta)\left(\frac{1}{2}(\sigma t - \theta/\sigma)^2 - \frac{\theta^2}{2\sigma^2}\right)^{\alpha-1} dt.$$

In the symmetric case $\theta = 0$ we find

$$\mu_A(dt) = \varphi_A(t)dt = \mu_B(dt) = \varphi_B(t)dt = \sigma^2 t\varphi\left(\frac{\sigma^2 t^2}{2}\right)dt = \frac{\alpha\sin(\alpha\pi)}{2^{\alpha-1}\pi}\sigma^{2\alpha}t^{2\alpha-1}dt,$$

i.e. $\sqrt{2}U_t/\sigma$ and $\sqrt{2}W_t/\sigma$ are stable processes of index 2α. Note that the skewness and kurtosis of G_t and Y_t are undefined. Figure 1 presents a simulation of the variance-stable process.

Variance product of stable processes

Here we take $G_1 = Z^{1/\alpha}X_\alpha$ where Z is a $\Gamma(\gamma, 1)$ random variable and X_α is a stable random variable with index $\alpha < 1$. The MGF of G_1 is $h(s) = (1 + s^\alpha)^\gamma$, cf. page 38 of [2], i.e. G_1 is a GGC with Thorin measure

$$\mu(dt) = \varphi(t)dt = \frac{1}{\pi}\frac{\gamma\alpha t^{\alpha-1}\sin(\alpha\pi)}{1 + t^{2\alpha} + 2t^\alpha\cos(\alpha\pi)}dt,$$

and Y_t decomposes as

$$Y_t = U_t - W_t,$$

where U_t and W_t are processes of independent product of stable increment and Thorin measures

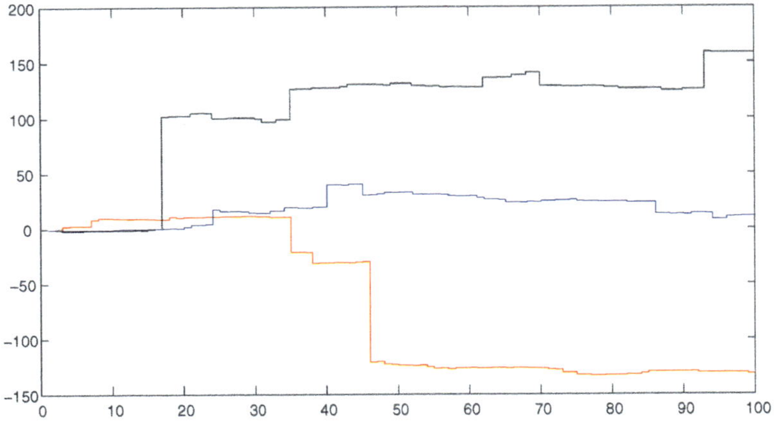

Fig. 2 Sample paths of variance-product of stable process with $\alpha = 0.99$ and $\gamma = 0.2$

$$\mu_A(dt) = \varphi_A(t)dt$$
$$= \frac{1}{\pi} \frac{\gamma\alpha((\sigma t + \theta/\sigma)^2/2 - \theta^2/(2\sigma^2))^{\alpha-1} \sin(\alpha\pi)(\sigma^2 t + \theta)}{1 + ((\sigma t + \theta/\sigma)^2/2 - \theta^2/(2\sigma^2))^{2\alpha} + 2((\sigma t + \theta/\sigma)^2/2 - \theta^2/(2\sigma^2))^{\alpha} \cos(\alpha\pi)} dt,$$

and

$$\mu_B(dt) = \varphi_B(t)dt$$
$$= \frac{1}{\pi} \frac{\gamma\alpha((\sigma t - \theta/\sigma)^2/2 - \theta^2/(2\sigma^2))^{\alpha-1} \sin(\alpha\pi)(\sigma^2 t - \theta)}{1 + ((\sigma t - \theta/\sigma)^2/2 - \theta^2/(2\sigma^2))^{2\alpha} + 2((\sigma t - \theta/\sigma)^2/2 - \theta^2/2\sigma^2)^{\alpha} \cos(\alpha\pi)} dt.$$

In the symmetric case

$$\mu_A(dt) = \varphi_A(t)dt = \mu_B(dt) = \varphi_B(t)dt$$
$$= \sigma^2 t \varphi\left(\frac{\sigma^2 t^2}{2}\right) dt = \frac{\gamma\alpha\sigma^{2\alpha} t^{2\alpha-1} \sin(\alpha\pi)}{\pi(2^{\alpha-1} + 2^{-\alpha-1}\sigma^{4\alpha} t^{4\alpha} + \sigma^{2\alpha} t^{2\alpha} \cos(\alpha\pi))} dt.$$

The skewness and kurtosis of G_t and Y_t are undefined. Figure 2 presents the corresponding simulation.

3 Sensitivity Analysis

In this section we extend approach of [8] to the sensitivity analysis of variance-GGC models. Consider $(B_t)_{t \in \mathbb{R}_+}$ a standard one-dimensional standard Brownian motion independent of the Lévy process $(Y_t)_{t \in [0,T]}$ generated by

$$Y_T := \int_0^\infty g(s)d\gamma_s.$$

Let Θ be a standard Gaussian random variable independent of $(Y_t)_{t\in[0,T]}$. For each $t \in [0, T]$, we denote by $\mathscr{F}_t$ the filtration generated by Θ and $\sigma(Y_s : s \in [0, t])$.

Let $(Z_t)_{t\in\mathbb{R}_+}$ be a real-valued stochastic process in $\mathbb{R}$ independent of $(Y_t)_{t\in\mathbb{R}_+}$ and $(B_t)_{t\in\mathbb{R}_+}$. Finally we denote by and let $C_b^n(\mathbb{R}_+;\mathbb{R})$ denote the class of n-time continuously differentiable functions with bounded derivatives, whereas $\mathscr{C}_c(\mathbb{R}_+;\mathbb{R})$ denotes the space of continuous functions with compact support.

Given $\theta \in \mathbb{R}$ and $\tau \in \mathbb{R}_+$ we consider the asset price S_T written as

$$S_T = S_0 \exp\left(\theta Y_T + \tau\sqrt{T}\Theta + Z_T + Tc(\theta,\tau)\right),$$

where the function $g(s) : \mathbb{R}_+ \to \mathbb{R}_+$ verifies (2).

Remark 1 When $\theta = 0$ the above model reduces to the standard Black-Scholes model, and in case $\theta \neq 0$ we find the variance-GGC model by taking $(Z_t)_{t\in[0,T]}$ to be a GGC process.

For example, we can take the Wiener-gamma integral $\int_0^\infty g(s)d\gamma_s$ to be a stable random variable and set Z_T to be another stable random variable, then the exponent of S_t is a variance-stable process. This example will be developed in the next section.

The next theorem deals with the sensitivity analysis of the variance-GGC model with respect to S_0, θ and τ, and is the main result in this section. Define the classes of functions

$$\mathscr{C}_L(\mathbb{R}_+;\mathbb{R}) := \{f \in C(\mathbb{R}_+;\mathbb{R}) \ : \ |f(x)| \le C(1+|x|) \text{ for some } C > 0\},$$

and

$$D(\mathbb{R}_+;\mathbb{R}) := \Big\{f : \mathbb{R}_+ \to \mathbb{R} \ : \ f = \sum_{k=1}^n c_k f_k \mathbf{1}_{A_k},\ n \ge 1,$$
$$c_k \in \mathbb{R},\ f_k \in \mathscr{C}_L(\mathbb{R}_+;\mathbb{R}),\ A_k \text{ intervals of } \mathbb{R}_+\Big\}.$$

Theorem 1 *Let $\Phi \in D(\mathbb{R}_+;\mathbb{R})$. Assume that the law of Z_T is absolutely continuous with respect to the Lebesgue measure, with*

$$\int_0^\infty \log\left(1 + \frac{g(s)f^k(s)}{(1-\lambda f(s))^{k+1}}\right) ds < \infty, \qquad k = 1, 2, 3. \tag{10}$$

Then

(i) (Delta—sensitivity with respect to S_0). We have

$$\frac{\partial}{\partial S_0}\mathbb{E}[\Phi(S_T)] = \frac{1}{S_0}\mathbb{E}[\Phi(S_T)L_T],$$

where

$$L_T = \frac{2\theta \int_0^\infty g(s) f^2(s) d\gamma_s}{(\theta \int_0^\infty g(s) f(s) d\gamma_s + \tau\sqrt{T}\eta)^2} + \frac{\int_0^\infty f(s) d\gamma_s - T \int_0^\infty f(s) ds + \eta\Theta}{\theta \int_0^\infty g(s) f(s) d\gamma_s + \tau\sqrt{T}\eta}.$$

(ii) (Sensitivity with respect to θ). We have

$$\frac{\partial}{\partial\theta}\mathbb{E}[\Phi(S_T)] = \mathbb{E}\left[\Phi(S_T)\left(L_T \int_0^\infty g(s) d\gamma_s - \frac{1}{H_T}\int_0^\infty g(s) f(s) d\gamma_s\right)\right] + T S_0 \frac{\partial c}{\partial\theta}(\theta, \tau)\frac{\partial}{\partial S_0}\mathbb{E}[\Phi(S_T)],$$

where $H_T = \theta \int_0^\infty g(s) f(s) d\gamma_s + \tau\sqrt{T}\eta.$

(iii) (Theta—sensitivity with respect to τ). We have

$$\frac{\partial}{\partial\tau}\mathbb{E}[\Phi(S_T)] = \mathbb{E}\left[\Phi(S_T) L_T \sqrt{T}\left(\Theta - \frac{\eta}{H_T}\right)\right] + T S_0 \frac{\partial c}{\partial\tau}(\theta, \tau)\frac{\partial}{\partial S_0}\mathbb{E}[\Phi(S_T)].$$

(iv) (Gamma—second derivative with respect to S_0). We have

$$\begin{aligned}
&\frac{\partial^2}{\partial S_0^2}\mathbb{E}[\Phi(S_T)] \\
&= \frac{1}{S_0^2}\mathbb{E}\left[\Phi(S_T)\left((L_T)^2 - \frac{1}{H_T}\left(\frac{I_T H_T - 2(K_T)^2}{(H_T)^3} + \frac{N_T H_T - M_T K_T}{(H_T)^2}\right)\right)\right] \\
&\quad - \frac{1}{S_0}\frac{\partial}{\partial S_0}\mathbb{E}[\Phi(S_T)],
\end{aligned}$$

where

$$K_T = 2\theta \int_0^\infty g(s) f^2(s) d\gamma_s, \quad M_T = \int_0^\infty f(s) d\gamma_s - T \int_0^\infty f(s) ds + \eta\Theta,$$

and

$$I_T = 6\theta \int_0^\infty g(s) f(s)^3 d\gamma_s, \quad N_T = \left(\int_0^\infty f(s) d\gamma_s - T \int_0^\infty f(s) ds + \eta\Theta\right)^2.$$

Next we state two lemmas which are needed for the proof of Theorem 1.

Lemma 1 *Assume that $\mathbb{E}[e^{2\gamma Z_T}] < \infty$ for some $\gamma > 1$. Let $f : \mathbb{R} \to (0, a)$ be a positive function and $\lambda \in (0, \varepsilon)$ for $\varepsilon < 1/a$ such that* (10) *holds. Fix $\eta > 0$ and suppose that one of the following conditions holds:*

(i) The density function of $Y_T = \int_0^\infty g(s)d\gamma_s$ *decays exponentially, or*
(ii) $\mathbb{E}\left[e^{2\gamma(1+\theta\delta)Y_T}\right] < \infty$ *for all* $\delta > 0$.

Let also

$$S_T^{(\lambda f)} = S_0 \exp\left(\theta \int_0^\infty \frac{g(s)}{1-\lambda f(s)} d\gamma_s + \tau\sqrt{T}(\Theta + \eta\lambda) + Z_T + c(\theta,\tau)T\right),$$

and

$$H_T^{(\lambda f)} = \frac{\partial}{\partial \lambda} \log S_T^{(\lambda f)} = \theta \int_0^\infty \frac{g(s)f(s)}{(1-\lambda f(s))^2} d\gamma_s + \tau\sqrt{T}\eta, \quad H_T = H_T^{(0)},$$

and

$$K_T^{(\lambda f)} = \frac{\partial}{\partial \lambda} H_T^{(\lambda f)} = 2\theta \int_0^\infty \frac{g(s)f^2(s)}{(1-\lambda f(s))^3} d\gamma_s, \quad K_T = K_T^{(0)}.$$

Then we have the $L^2(\Omega)$*-limits*

$$\lim_{\lambda\to 0} S_T^{(\lambda f)} H_T^{(\lambda f)} = S_T H_T \quad \text{and} \quad \lim_{\lambda \to 0} \frac{K_T^{(\lambda f)}}{(H_T^{(\lambda f)})^2} = \frac{K_T}{(H_T)^2}.$$

Proof For any $\lambda \in (0,\varepsilon)$, we have

$$\begin{aligned}
&\sup_{\lambda\in(0,\varepsilon)} \mathbb{E}\left[|S_T^{(\lambda f)} H_T^{(\lambda f)}|^{2\gamma}\right] \le C_1 \mathbb{E}\left[e^{2\gamma\tau\sqrt{T}\Theta}\right] \mathbb{E}\left[e^{2\gamma Z_T}\right] \\
&\quad \times \sup_{\lambda\in(0,\varepsilon)} \mathbb{E}\left[\left(\theta \int_0^\infty \frac{g(s)f(s)}{(1-\lambda f(s))^2} d\gamma_s + \tau\sqrt{T}\eta\right)^{2\gamma} \exp\left(2\gamma \int_0^\infty \frac{g(s)}{1-\lambda f(s)} d\gamma_s\right)\right] \\
&\le C_1 \mathbb{E}[e^{2\gamma\tau\sqrt{T}\Theta}] \mathbb{E}[e^{2\gamma Z_T}] \\
&\quad \times \sup_{\lambda\in(0,\varepsilon)} \left(\frac{a}{(1-\lambda a)^2}\right)^{2\gamma} \mathbb{E}\left[\left(\int_0^\infty g(s)d\gamma_s + \tau\sqrt{T}\eta\right)^{2\gamma} \exp\left(\frac{2\gamma\theta}{1-\lambda a} \int_0^\infty g(s)d\gamma_s\right)\right] \\
&\le C_1 \mathbb{E}[e^{2\gamma\tau\sqrt{T}\Theta}] \mathbb{E}[e^{2\gamma Z_T}] \\
&\quad \times \left(\frac{a}{(1-\varepsilon a)^2}\right)^{2\gamma} \mathbb{E}\left[\left(\int_0^\infty g(s)d\gamma_s + \tau\sqrt{T}\eta\right)^{2\gamma} \exp\left(\frac{2\gamma\theta}{1-\varepsilon a} \int_0^\infty g(s)d\gamma_s\right)\right],
\end{aligned}$$

where C_1 is a positive constant. Under condition (i) or (ii) above we have

$$\mathbb{E}\left[Y_T^{2\gamma} \exp\left(\frac{2\gamma\theta}{1-\varepsilon a} Y_T\right)\right] \le \mathbb{E}\left[\exp\left(2\gamma\left(1 + \frac{\theta}{1-\varepsilon a}\right) Y_T\right)\right] < \infty,$$

and similarly we have $\mathbb{E}\left[e^{\frac{2\gamma\theta}{1-\varepsilon a}Y_T}\right] < \infty$. Finally, we have $\mathbb{E}[e^{2\gamma Z_T}] < \infty$ by assumption, and it is clear that $\mathbb{E}[e^{2\gamma\tau\sqrt{T}\Theta}] < \infty$. Then $|S_T^{(\lambda f)} H_T^{(\lambda f)}|$ is $L^{2\gamma}(\Omega)$-integrable, hence $(S_T^{(\lambda f)} H_T^{(\lambda f)})^2$ is uniformly-integrable since $\gamma > 1$. Therefore, we have proved that $S_T^{(\lambda f)} H_T^{(\lambda f)}$ converges to $S_T H_T$ in $L^2(\Omega)$ as $\lambda \to 0$.

Next, for any $\lambda \in (0, \varepsilon)$ we have

$$\sup_{\lambda\in(0,\varepsilon)} \mathbb{E}[|K_T^{(\lambda f)}/(H_T^{(\lambda f)})^2|^{2\gamma}] \leq \sup_{\lambda\in(0,\varepsilon)} \mathbb{E}\left[\left(\left(\frac{2\theta}{\tau\sqrt{T}\eta}\right)\int_0^\infty \frac{g(s)f^2(s)}{(1-\lambda f(s))^3}d\gamma_s\right)^{2\gamma}\right]$$
$$\leq \left(\frac{a^2}{(1-\lambda a)^3}\right)^{2\gamma} \mathbb{E}\left[\left(\int_0^\infty g(s)d\gamma_s\right)^{2\gamma}\right] \sup_{\lambda\in(0,\varepsilon)}\left|\frac{2\theta}{\tau\sqrt{T}\eta}\right|^{2\gamma}$$
$$\leq \left(\frac{a^2}{(1-\varepsilon a)^3}\right)^{2\gamma} \mathbb{E}\left[\left(\int_0^\infty g(s)d\gamma_s\right)^{2\gamma}\right] \left|\frac{2\theta}{\tau\sqrt{T}\eta}\right|^{2\gamma},$$

since $\mathbb{E}\left[\left(\int_0^\infty g(s)d\gamma_s\right)^{2\gamma}\right]$ is finite under Condition (i) or (ii) above. Therefore $(K_T^{(\lambda f)}/(H_T^{(\lambda f)})^2)^2$ is uniformly-integrable since $\gamma > 1$, and this shows that $K_T^{(\lambda f)}/(H_T^{(\lambda f)})^2$ converges to $K_T/(H_T)^2$ as $\lambda \to 0$ in $L^2(\Omega)$. □

Lemma 2 *Assume that* $\mathbb{E}[e^{2\gamma Z_T}] < \infty$ *for some* $\gamma > 1$ *and that* (10) *holds. Suppose in addition that one of the following conditions holds:*

1. *The density function of* $\int_0^\infty g(s)d\gamma_s$ *decays exponentially.*
2. $\mathbb{E}\left[\left|e^{2\gamma(1+\theta\delta)Y_T}\right|\right] < \infty$ *for all* $\delta > 0$, *where* $Y_T = \int_0^\infty g(s)d\gamma_s$.

Then for $\Phi \in \mathscr{C}_b^1(\mathbb{R}_+, \mathbb{R})$ *it holds that*

(i) $$\mathbb{E}\left[\Phi'(S_T)S_T H_T\right] = \mathbb{E}\left[\left(\int_0^\infty f(s)d\gamma_s - T\int_0^\infty f(s)ds + \eta\Theta\right)\Phi(S_T)\right].$$

(ii) $\mathbb{E}[\Phi'(S_T)S_T] = \mathbb{E}[\Phi(S_T)L_T]$.

(iii) $$\mathbb{E}\left[\Phi'(S_T)S_T\int_0^\infty g(s)d\gamma_s\right] = \mathbb{E}\left[\Phi(S_T)\left(L_T\int_0^\infty g(s)d\gamma_s - \frac{1}{H_T}\int_0^\infty g(s)f(s)d\gamma_s\right)\right].$$

(iv) $$\mathbb{E}[\Phi'(S_T)S_T B_T] = \sqrt{T}\mathbb{E}\left[\Phi(S_T)L_T\left(\Theta - \frac{\eta}{H_T}\right)\right].$$

(v) If in addition $\Phi \in \mathscr{C}_b^2(\mathbb{R}_+, \mathbb{R})$ *and* (10) *is satisfied then we have*

$$\mathbb{E}[\Phi''(S_T)(S_T)^2] + \mathbb{E}[\Phi'(S_T)S_T]$$
$$= \mathbb{E}\left[\Phi(S_T)\left((L_T)^2 - \frac{1}{H_T}\left(\frac{I_T H_T - 2(K_T)^2}{(H_T)^3} + \frac{N_T H_T - M_T K_T}{(H_T)^2}\right)\right)\right].$$

Proof We have

$$\begin{aligned}\mathbb{E}[(\Phi(S_T))^2] &\leq 2\mathbb{E}[(\Phi(S_T)-\Phi(S_0))^2]+2\mathbb{E}[(\Phi(S_0))^2]\\ &\leq 2\mathbb{E}[(\Phi(S_0))^2]+2\int_0^1 \mathbb{E}[(\Phi'(rS_T+(1-r)S_0))^2(S_T-S_0)^2]dr\\ &< \infty,\end{aligned}$$

since $\Phi \in C_b^1(\mathbb{R}_+;\mathbb{R})$. As for (i) we have

$$\mathbb{E}\left[\Phi(S_T^{(\lambda f)})\right]=\mathbb{E}\left[\frac{dP_{\lambda f}}{dP}_{|\mathscr{F}_T}\Phi(S_T)\right], \tag{11}$$

where we define the probability measure $P_{\lambda f}$ via its Radon-Nikodym derivative

$$\frac{dP_{\lambda f}}{dP}_{|\mathscr{F}_T}=\frac{e^{\lambda\int_0^\infty f(s)d\gamma_s}}{\mathbb{E}[e^{\lambda\int_0^\infty f(s)d\gamma_s}]}\frac{e^{\lambda\eta\Theta}}{\mathbb{E}[e^{\lambda\eta\Theta}]}=e^{\lambda\int_0^\infty f(s)d\gamma_s+T\int_0^\infty \log(1-\lambda f(s))ds+\lambda\eta\Theta-\lambda^2\eta^2/2},$$

where $f:\mathbb{R}\to(0,a)$ and $\lambda\in(0,\varepsilon)$. In this way the GGC random variable $\int_0^\infty g(s)d\gamma_s$ and the Gaussian random variable Θ under $P_{\lambda f}$ are transformed to $\int_0^\infty \frac{g(s)}{1-\lambda f(s)}d\gamma_s$ and $\Theta+\eta\lambda$ under P.

First we prove that $\frac{\partial}{\partial\lambda}\mathbb{E}\left[\Phi(S_T^{(\lambda f)})\right]$ exists and equals the left hand side of (i). For every $\varepsilon\in(-\lambda,\lambda)$ we have

$$\frac{\Phi(S_T^{(\varepsilon f)})-\Phi(S_T)}{\varepsilon}=\int_0^1 \Phi'(S_T^{(r\varepsilon f)})S_T^{(r\varepsilon f)}H_T^{(r\varepsilon f)}dr,$$

and by the Cauchy-Schwarz inequality we get

$$\begin{aligned}&\mathbb{E}\left[\left|\frac{1}{\varepsilon}(\Phi(S_T^{(\varepsilon f)})-\Phi(S_T))-\Phi'(S_T)S_TH_T\right|\right]\\ &\quad\leq\int_0^1 \mathbb{E}[|\Phi'(S_T^{(r\varepsilon f)})S_T^{(r\varepsilon f)}H_T^{(r\varepsilon f)}-\Phi'(S_T)S_TH_T]dr\\ &\quad\leq\int_0^1\sqrt{\mathbb{E}[(\Phi'(S_T^{(r\varepsilon f)}))^2]}\sqrt{\mathbb{E}[(S_T^{(r\varepsilon f)}H_T^{(r\varepsilon f)}-S_TH_T)^2]}dr\\ &\qquad+\int_0^1\sqrt{\mathbb{E}[(\Phi'(S_T^{(r\varepsilon f)})-\Phi'(S_T))^2]}\sqrt{\mathbb{E}[(S_TH_T)^2]}dr.\end{aligned} \tag{12}$$

From the boundedness and continuity of $\Phi'(S_T^{(\varepsilon f)})$ with respect to ε in $L^2(\Omega)$, we have

$$\mathbb{E}[(\Phi'(S_T^{(\varepsilon f)}))^2] < \infty \quad \text{and} \quad \lim_{\varepsilon \to 0} \mathbb{E}[(\Phi'(S_T^{(\varepsilon f)}) - \Phi'(S_T))^2] = 0.$$

By Lemma 1 we get that $S_T^{(\lambda f)} H_T^{(\lambda f)}$ converges in $L^2(\Omega)$. Finally, we take the limit on both sides of (12) as $\varepsilon \to 0$. Next we prove that $\frac{\partial}{\partial \lambda} \mathbb{E}\left[\frac{dP_{\lambda f}}{dP}_{|\mathscr{F}_T} \Phi(S_T) \right]$ exists and equals the right hand side of (i).

For every $\varepsilon \in (-\lambda, \lambda)$ the Cauchy-Schwarz inequality yields

$$\begin{aligned}
&\mathbb{E}\left[\left| \frac{1}{\varepsilon}\left(\frac{dP_{\varepsilon f}}{dP}_{|\mathscr{F}_T} - \frac{dP_0}{dP}_{|\mathscr{F}_T} \right) \Phi(S_T) - \left(\int_0^\infty f(s) d\gamma_s - T \int_0^\infty f(s) ds + \eta \Theta \right) \Phi(S_T) \right| \right] \\
&\leq \sqrt{\mathbb{E}\left[(\Phi(S_T))^2 \right]} \\
&\quad \sqrt{\mathbb{E}\left[\left(\left(\frac{1}{\varepsilon}\left(\frac{dP_{\varepsilon f}}{dP}_{|\mathscr{F}_T} - \frac{dP_0}{dP}_{|\mathscr{F}_T} \right) \right) - \left(\int_0^\infty f(s) d\gamma_s - T \int_0^\infty f(s) ds + \eta \Theta \right) \right)^2 \right]}.
\end{aligned}$$

It is then straightforward to check that $\mathbb{E}[|\Phi(S_T)|^2] < \infty$ and

$$\frac{1}{\lambda}\left(\exp\left(\lambda \int_0^\infty f(s) d\gamma_s + T \int_0^\infty \log\left(1 - \lambda f(s)\right) ds + \lambda \eta \Theta - \lambda^2 \eta^2 / 2 \right) - 1 \right)$$

converges to

$$\int_0^\infty f(s) d\gamma_s - T \int_0^\infty f(s) ds + \eta \Theta$$

in $L^2(\Omega)$ as λ tends to 0 since $\lambda^{-1}(e^{\lambda \int_0^\infty f(s) d\gamma_s} - 1)$ converges to $\int_0^\infty f(s) d\gamma_s$ in $L^2(\Omega)$ as $\lambda \to 0$. We conclude by taking the limit on both sides as $\lambda \to 0$.

For (ii) we start with the identity

$$\mathbb{E}\left[\frac{\Phi(S_T^{(\lambda f)})}{H_T^{(\lambda f)}} \right] = \mathbb{E}\left[\frac{dP_{\lambda f}}{dP}_{|\mathscr{F}_T} \frac{\Phi(S_T)}{H_T} \right].$$

First we prove that $\frac{\partial}{\partial \lambda} \mathbb{E}\left[\frac{\Phi(S_T^{(\lambda f)})}{H_T^{(\lambda f)}} \right]$ exists and equals the left hand side of (ii). For every $\varepsilon \in [-\lambda, \lambda]$ we have

$$\frac{1}{\varepsilon}\left(\frac{\Phi(S_T^{(\varepsilon f)})}{H_T^{(\varepsilon f)}} - \frac{\Phi(S_T^{(0)})}{H_T} \right) = \int_0^1 \frac{\Phi'(S_T^{(r\varepsilon f)}) S_T^{(r\varepsilon f)} (H_T^{(r\varepsilon f)})^2 - \Phi(S_T^{(r\varepsilon f)}) K_T^{(r\varepsilon f)}}{(H_T^{(r\varepsilon f)})^2} dr,$$

and by the Cauchy-Schwarz inequality we get

$$\begin{aligned}
&\mathbb{E}\left[\left|\frac{1}{\varepsilon}\left(\frac{\Phi(S_T^{(\varepsilon f)})}{H_T^{(\varepsilon f)}}-\frac{\Phi(S_T)}{H_T}\right)-\frac{\Phi'(S_T)S_T(H_T)^2-\Phi(S_T)K_T}{(H_T)^2}\right|\right]\\
&\le \int_0^1 \mathbb{E}\left[\left|\frac{\Phi'(S_T^{(r\varepsilon f)})S_T^{(r\varepsilon f)}(H_T^{(r\varepsilon f)})^2-\Phi(S_T^{(r\varepsilon f)})K_T^{(r\varepsilon f)}}{(H_T^{(r\varepsilon f)})^2}-\frac{\Phi'(S_T)S_T^{(0)}(H_T)^2-\Phi(S_T)K_T}{(H_T)^2}\right|\right]dr\\
&\le \int_0^1 \sqrt{\mathbb{E}[(\Phi'(S_T^{(r\varepsilon f)}))^2]}\sqrt{\mathbb{E}[(S_T^{(r\varepsilon f)}-S_T)^2]}dr\\
&+\int_0^1 \sqrt{\mathbb{E}[(\Phi'(S_T^{(r\varepsilon f)})-\Phi'(S_T))^2]}\sqrt{\mathbb{E}[(S_T)^2]}dr\\
&+\int_0^1 \sqrt{\mathbb{E}[(\Phi(S_T^{(r\varepsilon f)}))^2]}\sqrt{\mathbb{E}[(K_T^{(r\varepsilon f)}/(H_T^{(r\varepsilon f)})^2-K_T/(H_T)^2)^2]}dr\\
&+\int_0^1 \sqrt{\mathbb{E}[(\Phi(S_T^{(r\varepsilon f)})-\Phi(S_T))^2]}\sqrt{\mathbb{E}[(K_T/(H_T)^2)^2]}dr. \qquad (13)
\end{aligned}$$

We have shown $\mathbb{E}[(\Phi(S_T))^2]<\infty$ in the proof of (i). Then

$$\begin{aligned}
\mathbb{E}[(\Phi(S_T^{(\varepsilon f)}))^2] &\le 2\mathbb{E}[(\Phi(S_T^{(\varepsilon f)})-S_T)^2]+2\mathbb{E}[(\Phi(S_T))^2]\\
&\le 2\varepsilon^2\int_0^1 \mathbb{E}[(\Phi'(S_T^{(r\varepsilon f)})S_T^{(r\varepsilon f)}H_T^{(r\varepsilon f)})^2]dr+2\mathbb{E}[(\Phi(S_T))^2]\\
&\le 2\varepsilon^2\sup_{x\in\mathbb{R}}|\Phi'(x)|^2\sup_{|\varepsilon|\le\lambda}\mathbb{E}[(S_T^{(\varepsilon f)}H_T^{(\varepsilon f)})^2]+2\mathbb{E}[(\Phi(S_T))^2]<\infty,
\end{aligned}$$

where the Cauchy-Schwarz inequality and the Fubini theorem have been used for the second inequality. The convergence of $S_T^{(\varepsilon f)}H_T^{(\varepsilon f)}$ as $\varepsilon\to 0$ in $L^2(\Omega)$ has been proved in Lemma 1. Note that $\mathbb{E}[(\Phi(S_T^{(\varepsilon f)}))^2]<\infty$ also implies

$$\mathbb{E}[(\Phi(S_T^{(\varepsilon f)})-\Phi(S_T))^2]\to 0 \quad as \quad \varepsilon\to 0.$$

By Lemma 1, we get $K_T^{(\varepsilon f)}/(H_T^{(\varepsilon f)})^2$ converges to $K_T/(H_T)^2$ as $\varepsilon\to 0$ in $L^2(\Omega)$. Taking the limit on both sides of (13) as $\varepsilon\to 0$.

Next, we prove that $\frac{\partial}{\partial\lambda}\mathbb{E}\left[\frac{dP_{\lambda f}}{dP}\Big|_{\mathscr{F}_T}\frac{\Phi(S_T)}{H_T}\right]$ exists and is equal to the right hand side of (ii). For all $p>0$ we have

$$\mathbb{E}[(H_T^{(\lambda f)})^{-2p}]=\int_0^\infty\left(\theta\int_0^\infty\frac{g(s)f(s)}{(1-\lambda f(s))^2}d\gamma_s+\tau\sqrt{T}\eta\right)^{-2p}f_1(y)dy<(\tau\sqrt{T}\eta)^{-2p},$$

where f_1 is the density function of $\int_0^\infty\frac{g(s)f(s)}{(1-\lambda f(s))^2}d\gamma_s$. Therefore, the moment is uniformly bounded.

We conclude as in the second part of proof of (i). The proof of $(iii)-(iv)$ is similar to that of (ii). As for (iii) we have

$$\mathbb{E}\left[\frac{\Phi(S_T^{(\lambda f)})}{H_T^{(\lambda f)}}\int_0^\infty \frac{g(s)}{1-\lambda f(s)}d\gamma_s\right] = \mathbb{E}\left[\frac{dP_{\lambda f}}{dP}_{|\mathscr{F}_T}\frac{\Phi(S_T)}{H_T}\int_0^\infty g(s)d\gamma_s\right].$$

For the first part, the existence of the derivative can be obtained as

$$\begin{aligned}\mathbb{E}\left[\left(\int_0^\infty \frac{g(s)}{1-\lambda f(s)}d\gamma_s - \int_0^\infty g(s)d\gamma_s\right)^2\right] &\leq \mathbb{E}\left[\left(\lambda\int_0^\infty g(s)\frac{f(s)}{1-\lambda f(s)}d\gamma_s\right)^2\right]\\ &\leq \mathbb{E}\left[\left(\lambda\frac{a}{1-\lambda a}\int_0^\infty g(s)d\gamma_s\right)^2\right]\\ &\leq \infty.\end{aligned}$$

Similarly, $\int_0^\infty \frac{g(s)f(s)}{(1-\lambda f(s))^2}d\gamma_s$ converges to $\int_0^\infty g(s)f(s)d\gamma_s$ in $L^2(\Omega)$ as $\lambda \to 0$. The second part is almost the same as (i) by uniform boundedness of $H_T^{(\lambda f)}$.

For (iv) we have

$$(\Theta + \eta\lambda)\mathbb{E}\left[\frac{\Phi(S_T^{(\lambda f)})}{H_T^{(\lambda f)}}\right] = \Theta\mathbb{E}\left[\frac{dP_{\lambda f}}{dP}_{|\mathscr{F}_T}\frac{\Phi(S_T)}{H_T}\right].$$

For the first part, the existence of the derivative follows from the fact that Θ has a Gaussian distribution. The second part is proved similarly.

Finally, for (v), define $\Psi(x) = \Phi'(x)x$, and by the result of (ii) we have

$$\mathbb{E}[\Phi''(S_T)(S_T)^2] = \mathbb{E}[(\Psi'(S_T) - \Phi'(S_T))S_T] = \mathbb{E}[\Psi(S_T)L_T] - \mathbb{E}[\Phi'(S_T)S_T].$$

Hence, we obtain the desired equation by differentiating

$$\mathbb{E}\left[\Phi(S_T^{(\lambda f)})\frac{L_T^{(\lambda f)}}{H_T^{(\lambda f)}}\right] = \mathbb{E}\left[\frac{dP_{\lambda f}}{dP}_{|\mathscr{F}_T}\Phi(S_T)\frac{L_T}{H_T}\right]$$

at $\lambda = 0$. □

Now we can prove Theorem 1.

Proof The proof of Theorem 1 uses the same argument as in the proof of Corollary 3.6 of [9]. The only difference is that S_T is a variance-gamma process in the proof of Corollary 3.6 of [9], while S_T is a variance-GGC process in this proof.

When $\Phi \in \mathscr{C}_b^2(\mathbb{R}_+, \mathbb{R})$, all four formulas in Theorem 1 are direct consequences of $(ii) - (v)$ in Lemma 2, and we now extend this result to the class $D(\mathbb{R}_+; \mathbb{R})$. In general, in order to obtain an extension to Φ in a class $\Re_1$ of functions based on an approximating sequence $(\Phi_n)_{n\in\mathbb{N}}$ in a class $\Re_2 \subset \Re_1$, it suffices to show that for each compact set $K \subset \mathbb{R}$ we have

$$\sup_{S_0 \in K} |\mathbb{E}[\Phi_n(S_T)] - \mathbb{E}[\Phi(S_T)]| \to 0 \quad as \quad n \to \infty, \tag{14}$$

and

$$\lim_{n\to\infty} \sup_{S_0 \in K} \left| \frac{\partial}{\partial S_0} \mathbb{E}[\Phi_n(S_T)] - \frac{1}{S_0} \mathbb{E}[\Phi(S_T) L_T] \right| = 0. \tag{15}$$

The extension is then based on the above steps, first from $\mathscr{C}_b^2(\mathbb{R}_+, \mathbb{R})$ to $\mathscr{C}_c(\mathbb{R}_+, \mathbb{R})$, then to $\mathscr{C}_b(\mathbb{R}_+, \mathbb{R})$ and to the class of finite linear combinations of indicator functions on an interval of $\mathbb{R}$. Finally the result is extended to the class of functions Φ of the form $\Phi = \Psi \times \mathbf{1}_A$ where $\Psi \in \mathscr{C}_L(\mathbb{R}_+, \mathbb{R})$ and A is an interval of $\mathbb{R}_+$. This shows that (14) and (15) are satisfied, and the details of each step are the same as in the proof of Corollary 3.6 of [9]. □

Acknowledgments This research was supported by MOE Tier 2 grant MOE2012-T2-2-033 ARC 3/13.

References

1. Barndorff-Nielsen, O.E.: Processes of normal inverse Gaussian type. Finance Stochast. **2**(1), 41–68 (1998)
2. Bondesson, L.: Generalized gamma convolutions and related classes of distributions and densities. Lecture Notes in Statistics, vol. 76. Springer, New York (1992)
3. Boyarchenko, S.I., Levendorskiĭ, S.Z.: Non-Gaussian Merton-Black-Scholes theory. Advanced Series on Statistical Science & Applied Probability, vol. 9. World Scientific Publishing Co., Inc., River Edge, NJ (2002)
4. Carr, P.P., Geman, H., Madan, D.B., Yor., M.: The fine structure of asset returns: an empirical investigation. J. Business, **75**(2), (2002)
5. Cox, J.C., and Ross, S.A.: The valuation of options for alternative stochastic processes. J. Financ. Econ, **3**, 145–166 (1976)
6. Geman, H., Madan, D.B., Yor, M.: Asset prices are Brownian motion: only in business time. In: Quantitative analysis in financial markets, pp. 103–146. World Scientific Publishing, River Edge, NJ (2001)
7. James, L.F., Roynette, B., Yor, M.: Generalized gamma convolutions, Dirichlet means, Thorin measures, with explicit examples. Prob. Surv. **5**, 346–415 (2008)
8. Kawai, R., Kohatsu-Higa, A.: Computation of Greeks and multidimensional density estimation for asset price models with time-changed brownian motion. Appl. Math. Finance **17**(4), 301–321 (2010)
9. Kawai, R., Takeuchi, A.: Greeks formulas for an asset price model with gamma processes. Math. Finance **21**(4), 723–742 (2011)
10. Kawai, R., Takeuchi, A.: Computation of Greeks for asset price dynamics driven by stable and tempered stable processes. Quant. Finance **13**(8), 1303–1316 (2013)
11. Madan, D.B., Carr, P.P., Chang, E.C.: The variance gamma process and option pricing. Eur. Finance Rev. **2**, 79–105 (1998)
12. Madan, D.B., Seneta, E.: The variance gamma model for share market returns. J. Bus. **63**, 511–524 (1990)
13. Merton, R.: Option pricing when underlying stock returns are discontinuous. J. Financ. Econ. **3**, (1976)
14. Privault, N., Yang, D.C.: Infinite divisibility of interpolated gamma powers. J. Math. Anal.Appl. **405**(2), 373–387 (2013)

Decomposition of the Pricing Formula for Stochastic Volatility Models Based on Malliavin-Skorohod Type Calculus

Josep Vives

Abstract The goal of this survey article is to present in detail a method that, for a financial derivative under a certain stochastic volatility model, allows to obtain a decomposition of its pricing formula that distinguishes clearly the impact of correlation and jumps. This decomposed pricing formula, usually called Hull and White type formula, can be potentially useful for model selection and calibration. The method is based on the obtention of an ad-hoc anticipating Itô formula.

Keywords Hull and White type formula · Malliavin-Skorohod calculus · Stochastic volatility jump-diffusion models · Derivative pricing · Quantitative finance

Mathematical Subject Classification 60H07 · 60H30 · 91G80 · 91G20

1 Introduction

The decomposition method presented in this paper is based on a series of works developed during the last ten years. In [1], E. Alòs obtained a decomposition of the pricing formula, usually called Hull and White type formula, for a plain vanilla call under a correlated stochastic volatility model, with minor hypothesis on the volatility process related with its Malliavin derivability. The decomposition was obtained applying an ad-hoc extension of the anticipative Itô formula given in [2]. The obtained formula showed clearly the impact on prices of adding correlation between price and volatility in stochastic volatility models.

In [3] the same type of formula was obtained adding also finite activity jumps in the price process. A new term appeared, showing the impact of jumps. In [5] the for-

Supported by grant MEC MTM 2012-31192.

J. Vives (✉)
Department of Mathematics and Computer Science,
Institute of Mathematics of the University of Barcelona (IMUB),
University of Barcelona, Barcelona, Spain
e-mail: josep.vives@ub.edu

M. Eddahbi et al. (eds.), *Statistical Methods and Applications in Insurance and Finance*, Springer Proceedings in Mathematics & Statistics 158,
DOI 10.1007/978-3-319-30417-5_4

mula was extended to the case of assuming jumps also on the volatility process. Still a new term appeared in the formula. Finally, on [9], the result in [5] was extended for free to the case of infinite activity and finite variation jumps, and with a certain restriction in the interpretation of the formula, to the case of infinite activity and infinite variation jumps. The very general model considered in this last paper covers almost all stochastic volatility models with and without jumps, treated in the literature.

As we see in the paper, the presence of correlation and jumps in stochastic volatility models is relevant. Additional terms in the pricing formula appears from correlation, from jumps in the price process and from jumps in the volatility. Malliavin-Skorohod calculus and the decomposition method allow to obtain these pricing formulas that clearly distinguish the effect of correlation than the effect of jumps, for different types of jump models. If the stochastic volatility is correlated only with the continuous part of the price process, only Gaussian Malliavin-Skorohod calculus is needed. If the stochastic volatility is also correlated with price jumps, Lévy Malliavin-Skorohod calculus is needed.

Section 2 is devoted to the Brownian (no jump) case and Sect. 3 treats the Lévy case.

2 Decomposition of the Pricing Formula Under a General Brownian Stochastic Volatility Model

The main reference of the theory presented in this section is [1].

2.1 The Model

Let $T > 0$ be a finite horizon, $S = \{S_t, t \in [0, T]\}$ a price process, $X_t = \log S_t$ the corresponding log price process and $r > 0$ the fixed interest rate. We assume the following exponential model with stochastic volatility for the dynamics of the log-price, under a market chosen risk-neutral probability:

$$X_t = x + rt - \frac{1}{2}\int_0^t \sigma_s^2 ds + \int_0^t \sigma_s (\rho dW_s + \sqrt{1-\rho^2} dB_s)$$

where x is the current log-price, W and B are independent standard Brownian motions and $\rho \in (-1, 1)$.

We denote by $\mathscr{F}^W$ and $\mathscr{F}^B$ the filtrations generated by the independent processes W and B. Moreover, we define $\mathscr{F}$, the filtration associated to S, by $\mathscr{F} := \mathscr{F}^W \vee \mathscr{F}^B$. We consider our price model defined on the product of the canonical spaces of processes W and B.

The volatility process σ is assumed to be a square-integrable stochastic process, adapted to $\mathscr{F}^W$ and with strictly positive and càdlàg trajectories.

Note that this is a very general stochastic volatility model. In this sense, recall the following facts:

- The model is a generalization of Heston model or other classical correlated stochastic volatility models in the sense that we do not assume a concrete dynamics for the volatility process σ.
- If $\rho = 0$ we have a generalization in the same sense as before of different non correlated stochastic volatility models as Hull-White, Scott, Stein-Stein or Ball-Roma.
- If σ is deterministic or constant we have the classical Osborne-Samuelson-Black-Scholes model.

For information about correlated and non-correlated stochastic volatility models, a good reference is [8].

Stochastic volatility models pursue the goal to replicate price surfaces of plain vanilla options (depending on time to maturity and strike) given by derivative markets or vanilla desks. The stochastic volatility σ is a process not directly observable, so it is not easy to model. This is a justification for trying to assume minimal conditions on it.

Let H_T be the payoff of a financial derivative. Assume it is a $\mathscr{F}_T$−measurable functional. Its price is given by $V_t = e^{-r(T-t)}\mathbb{E}_t(H_T)$ where $\mathbb{E}_t := \mathbb{E}(\cdot|\mathscr{F}_t)$. To fix ideas we will concentrate on the case of a plain vanilla call, that is, $H_T = (S_T - K)^+$. So, our goal is to obtain a decomposition of

$$V_t = e^{-r(T-t)}\mathbb{E}_t((S_T - K)^+)$$

under our risk neutral model, in order to clarify the effect of correlation in the price.

2.2 *Fast Summary of Brownian Malliavin-Skorohod Calculus*

Here we simply recall some basic definitions and facts necessary for our purpose. See for example [10] for a complete presentation of the theory.

Let W and $(\Omega^W, \mathscr{F}^W, \mathbb{P}_W)$ be the canonical Wiener process and its canonical space, respectively. Recall that $\Omega^W := C_0([0, T])$ is the space of continuous functions on $[0, T]$, null at the origin. Denote by $\mathbb{E}_W$ the expectation with respect to $\mathbb{P}_W$.

Consider the family of smooth functionals of type

$$F = f(W_{t_1}, \ldots, W_{t_n})$$

for any $n \geq 0$, $t_1, \ldots t_n \in [0, T]$ and $f \in C_b^\infty(\mathbb{R}^n)$.

Given a smooth functional F we define its Malliavin derivative $D^W F$ as the element of $L^2(\Omega^W \times [0, T])$ given by

$$D_t F = \sum_{i=1}^{n} \partial_i f(W_{t_1}, \ldots, W_{t_n}) 1\!\!1_{[0,t_i]}(t).$$

The operator D^W is closed and densely defined in $L^2(\Omega^W)$, and its domain $Dom\ D^W$ is the closure of the smooth functionals with respect the norm

$$||F||_{Dom\ D^W} := \left(\mathbb{E}_W(|F|^2) + \mathbb{E}_W \int_0^T |D_t^W F|^2 dt\right)^{\frac{1}{2}}.$$

We define δ^W as the dual operator of D^W. Given $u \in L^2(\Omega^W \times [0, T])$, $\delta^W(u)$ is the element of $L^2(\Omega^W)$ characterized by

$$\mathbb{E}_W(F\delta^W(u)) = \mathbb{E}_W \int_0^T u_t D_t^W F dt$$

for any $F \in Dom\ D^W$. Note that taking $F \equiv 1$ we obtain

$$\mathbb{E}_W(\delta^W(u)) = 0.$$

The following results will be helpful:

- If F, G and $F \cdot G$ belong to $Dom\ D^W$ we have

$$D^W(F \cdot G) = F D^W G + G D^W F.$$

- If $F \in Dom\ D^W$, $u \in Dom\ \delta^W$ and $F \cdot u \in Dom\ \delta^W$ then

$$\delta^W(F \cdot u) = F\delta^W(u) - \int_0^T u_t D_t^W F dt.$$

- It is well known that D^W can be interpreted as a directional derivative on the Wiener space and δ^W is an extension of the classical Itô integral.

We define the space $\mathbb{L}_W^{1,2} := L^2([0, T]; Dom\ D^W)$, that is the space of processes $u \in L^2([0, T] \times \Omega^W)$ such that $u_t \in Dom\ D^W$ for almost all t and $Du \in L^2(\Omega^W \times [0, T]^2)$. It can be proved that $\mathbb{L}_W^{1,2} \subseteq Dom\ \delta^W$ and

$$\mathbb{E}_W(\delta^W(u)^2) \leq ||u||^2_{\mathbb{L}_W^{1,2}} := \mathbb{E}_W(||u||^2_{L^2([0,T])}) + \mathbb{E}_W(||D^W u||^2_{L^2([0,T]^2)}).$$

Finally, we will denote $\delta_t^W(u) := \delta^W(u 1\!\!1_{[0,t]})$.

2.3 The Hull and White Formula

If we assume constant volatility we have the well known geometric Brownian model. In this case, the price V_t is given by the well known Black-Scholes formula:

$$V_t = BS(t, X_t, \sigma) = e^x \Phi(d_+) - Ke^{-r(T-t)}\Phi(d_-)$$

where

$$d_\pm = \frac{X_t - \log K + r(T-t)}{\sigma\sqrt{T-t}} \pm \frac{\sigma\sqrt{T-t}}{2}$$

and Φ is the cumulative probability function of the standard normal law.

If we allow $\sigma = \sigma(t)$ be a deterministic function, it is easy to see, that

$$X_T - X_t \sim \mathcal{N}((r - \frac{1}{2}\bar{\sigma}_t^2)(T-t), \bar{\sigma}_t^2(T-t)),$$

where

$$\bar{\sigma}_t := \sqrt{\frac{1}{T-t}\int_t^T \sigma^2(s)ds}$$

is the so called future average volatility. Define $\bar{\sigma}_T$ as the limit of $\bar{\sigma}_t$ when $t \uparrow T$.

So, in this case, the pricing formula is exactly the Black-Scholes formula changing σ by $\bar{\sigma}_t$, that is, $V_t = BS(t, X_t, \bar{\sigma}_t)$. This suggests that it is the future average volatility and not the volatility the really relevant quantity in pricing. Black-Scholes formula would be nothing more than the particular case of constant future average volatility.

If σ is a stochastic process uncorrelated with price, that is, $\rho = 0$ in our model, we have, following for example [7]:

$$V_t = \mathbb{E}_t(BS(t, X_t, \bar{\sigma}_t)).$$

This is the classical Hull and White formula and covers non correlated stochastic volatility models as the cases of Hull-White, Scott, Stein-Stein, Ball-Roma and others. The proof is immediate, conditioning first by $\mathcal{F}_t \vee \mathcal{F}_T^W$.

Note that the future average volatility $\bar{\sigma}_t$ is an anticipative process. This suggest the use of Malliavin-Skorohod calculus as a natural tool to deal with this type of processes.

In the correlated case we have the following theorem:

Theorem 1 *Assume*

- *(A1):* $\sigma^2 \in \mathbb{L}_W^{1,2}$.
- *(A2):* $\sigma \in \mathbb{L}_W^{1,2}$.

Then we have,

$$V_t = \mathbb{E}_t[BS(t, X_t, \bar{\sigma}_t)] + \frac{\rho}{2}\mathbb{E}_t[\int_t^T e^{-r(s-t)}(\partial_x^3 - \partial_x^2)BS(s, X_s, \bar{\sigma}_s)\Lambda_s ds]$$

where

$$\Lambda_s := (\int_s^T D_s^W \sigma_r^2 dr)\sigma_s.$$

Proof The proof is based on the so called *decomposition method.*

Recall that

$$V_T = (e^{X_T} - K)^+ = BS(T, X_T, \bar{\sigma}_T)$$

and so

$$e^{-rt}V_t = \mathbb{E}_t(e^{-rT}BS(T, X_T, \bar{\sigma}_T)).$$

The idea of the proof consists in applying an ad-hoc anticipative Itô formula to the process

$$e^{-rs}BS(s, X_s, \bar{\sigma}_s)$$

between t and T, take conditional expectations $\mathbb{E}_t$ and multiply by e^{rt}. This gives the expansion for V_t.

The ad-hoc Itô formula is an adaptation to our case of the anticipative Itô formula proved in [2]. Define

$$Y_t := (T-t)\bar{\sigma}_t^2 = \int_t^T \sigma_r^2 dr.$$

Thanks to (A1), we are under the conditions of Theorem 1 in [1], and so, for any $F \in C_b^{1,2,2}([0,T] \times \mathbb{R} \times [0,\infty))$, we have

$$\begin{aligned} F(t, X_t, Y_t) &= F(0, X_0, Y_0) + \int_0^t \partial_s F(s, X_s, Y_s)ds + \delta_t^{W,B}(\partial_x F(\cdot, X_\cdot, Y_\cdot)\sigma_\cdot) \\ &+ \int_0^t \partial_x F(s, X_s, Y_s)(r - \frac{\sigma_s^2}{2})ds - \int_0^t \partial_y F(s, X_s, Y_s)\sigma_s^2 ds \\ &+\rho \int_0^t \partial_{xy} F(s, X_s, Y_s)\Lambda_s ds + \frac{1}{2}\int_0^t \partial_x^2 F(s, X_s, Y_s)\sigma_s^2 ds. \end{aligned}$$

Now we want to apply this result to

$$F(s, x, y) := e^{-rs}BS(s, x, \sqrt{\frac{y}{T-s}}),$$

but this function doesn't satisfy the required conditions of the previous Itô formula because the derivatives are not bounded, so we need to use a mollifier argument.

For $n \geq 1$ and $\delta > 0$, we consider the approximation,

$$F_{n,\delta}(s, x, y) := e^{-rs} BS(s, x, \sqrt{\frac{y+\delta}{T-s}})\phi(\frac{x}{n}),$$

where $\phi \in C_b^2(\mathbb{R})$, such that $\phi(z) = 1$ if $|z| \leq 1$, $\phi(z) \in [0, 1]$ if $|z| \in [1, 2]$ and $\phi(z) = 0$ if $|z| > 2$.

Then, applying the previous ad-hoc Itô formula to $F_{n,\delta}(s, X_s, Y_s)$, taking the conditional expectation $\mathbb{E}_t$, using the fact that Skorohod type integrals have zero expectation and multiplying by e^{rt} we obtain

$$\begin{aligned}
&\mathbb{E}_t(e^{-r(T-t)} BS(T, X_T, \bar{\sigma}_T^\delta)\phi(\frac{X_T}{n})) \\
= \ &\mathbb{E}_t(BS(t, X_t, \bar{\sigma}_t^\delta)\phi(\frac{X_t}{n})) \\
&+\mathbb{E}_t(\int_t^T e^{-r(s-t)} A_n(s)ds) \\
&+\frac{\rho}{2}\mathbb{E}_t(\int_t^T e^{-r(s-t)}(\partial_x^3 - \partial_x^2)BS(s, X_s, \bar{\sigma}_s^\delta)\phi_n(\frac{X_s}{n})\Lambda_s ds \\
&+\frac{\rho}{2}\mathbb{E}_t(\int_t^T e^{-r(s-t)}(\partial_x^2 - \partial_x)BS(s, X_s, \bar{\sigma}_s^\delta)\frac{1}{n}\phi'(\frac{X_s}{n}))\Lambda_s ds
\end{aligned}$$

where

$$\bar{\sigma}_s^\delta := \sqrt{\frac{Y_s + \delta}{T-s}}$$

and

$$\begin{aligned}
A_n(s) : = &\frac{\sigma_s^2}{n}\partial_x BS(s, X_s, \bar{\sigma}_s^\delta)\phi'(\frac{X_s}{n}) \\
&+\frac{\sigma_s^2}{2n} BS(s, X_s, \bar{\sigma}_s^\delta)(\frac{1}{n}\phi''(\frac{X_s}{n}) - \phi'(\frac{X_s}{n})) \\
&+\frac{r}{n} BS(s, X_s, \bar{\sigma}_s^\delta)\phi'(\frac{X_s}{n}).
\end{aligned}$$

The details can be found in [9] (erasing there the terms depending on jumps, that will be treated later in this paper).

Finally, the result follows from the dominated convergence theorem taking limits first on $n \uparrow \infty$ and then on $\delta \downarrow 0$. The dominated convergence runs thanks to the properties of Black-Scholes function and (A2). For the left hand side and the two first terms on the right hand side we use the fact that function $BS(t, x, \sigma)$ is bounded by $e^x + K$ and its derivative $(\partial_x BS)(t, x, \sigma)$ is bounded by e^x. For the last two terms on the right hand side we use Lemma 2 in [3] that says that for any $n \geq 0$,

$$|\mathbb{E}[\partial_x^n(\partial_x^2 - \partial_x)BS(s, X_s, \bar{\sigma}_s)|\mathscr{F}_t \vee \mathscr{F}_T^W]| \leq C_n(\rho)(\int_t^T \sigma_s^2 ds)^{-\frac{n+1}{2}},$$

for a certain constant $C_n(\rho)$ that depends only on n and ρ.

For example, for the third term on the right hand side, we have

$$\mathbb{E}_t(\int_t^T e^{-r(s-t)}(\partial_x^3 - \partial_x^2)BS(s, X_s, \bar{\sigma}_s^\delta)\phi_n(\frac{X_s}{n})\Lambda_s ds)$$
$$= \mathbb{E}_t(\int_t^T e^{-r(s-t)}\mathbb{E}[(\partial_x^3 - \partial_x^2)BS(s, X_s, \bar{\sigma}_s^\delta)|\mathscr{F}_t \vee \mathscr{F}_T^W]\phi_n(\frac{X_s}{n})\Lambda_s ds).$$

And, applying the lemma,

$$|\mathbb{E}[(\partial_x^3 - \partial_x^2)BS(s, X_s, v_s^\delta)|\mathscr{F}_t \vee \mathscr{F}_T^W]\phi_n(\frac{X_s}{n})\Lambda_s| \leq C_1(\rho)\frac{|\Lambda_s|}{\int_t^T \sigma_s^2 ds}.$$

Using the chain rule for D^W, the problem reduces to show

$$\mathbb{E}_t(\frac{\int_t^T(\int_t^T |D_s^W\sigma_u|\sigma_u dr)\sigma_s ds}{\int_t^T \sigma_r^2 dr}) < \infty,$$

and applying Cauchy-Schwarz inequality twice, we can bound this expression by

$$\mathbb{E}_t(\frac{(\int_t^T(\int_t^T |D_s^W\sigma_u|\sigma_u du)^2 ds)^{\frac{1}{2}}}{(\int_t^T \sigma_u^2 du)^{\frac{1}{2}}})$$
$$\leq \mathbb{E}_t(\frac{(\int_t^T(\int_t^T |D_s^W\sigma_u|^2 dr)(\int_t^T \sigma_u^2 du)ds)^{\frac{1}{2}}}{(\int_t^T \sigma_u^2 du)^{\frac{1}{2}}})$$
$$\leq \mathbb{E}_t((\int_t^T \int_t^T |D_s^W\sigma_u|^2 duds)^{\frac{1}{2}})$$
$$\leq (\mathbb{E}_t \int_t^T \int_t^T |D_s^W\sigma_u|^2 duds)^{\frac{1}{2}}.$$

So, (A2) proves that this expression is finite.

For the fourth term in the right hand side, applying the lemma and using C as a generic constant, we have

$$|\mathbb{E}[(\partial_x^2 - \partial_x)BS(s, X_s, \bar{\sigma}_s^\delta)|\mathscr{F}_t \vee \mathscr{F}_T^W]\frac{1}{n}\phi_n'(\frac{X_s}{n})\Lambda_s| \leq \frac{C}{n}\frac{|\Lambda_s|}{(\int_t^T \sigma_s^2 ds)^{\frac{1}{2}}}.$$

So, we have to show

$$\mathbb{E}_t\left(\frac{\int_t^T \sigma_s(\int_s^T |D_s^W \sigma_u|\sigma_u du)ds}{(\int_t^T \sigma_u^2 du)^{\frac{1}{2}}}\right) < \infty,$$

that follows applying Cauchy-Schwarz inequality, similarly to the previous case.

Remark 1 Note that hypothesis (A2) can be changed by the following alternative hypothesis of uniform ellipticity (A2′): The process σ^2 defined on $[0, T]$ is uniformly bounded below by a constant $a > 0$. In fact (A1) and (A2′), jointly, imply (A2).

3 The Lévy Case

The main references for this section are [3, 5, 9].

3.1 A Very General Stochastic Volatility Lévy Model

Assume now the following exponential Lévy model with stochastic volatility for the dynamics of the log-price, under a market chosen risk-neutral probability:

$$X_t = x + rt - \frac{1}{2}\int_0^t \sigma_s^2 ds + \int_0^t \sigma_s(\rho dW_s + \sqrt{1-\rho^2}dB_s) + L_t^0$$

where L^0 is a pure jump Lévy process with possibly infinitely many jumps with triplet $(\gamma_0, 0, \nu)$ and independent of W and B. Now, the volatility process σ is assumed to be adapted to the filtration generated by W and L^0.

Due to the well known Lévy-Itô decomposition we can write

$$L_t^0 = \gamma_0 t + \int_0^t \int_{\{|y|>1\}} yN(ds, dy) + \lim_{\varepsilon \downarrow 0} \int_0^t \int_{\{\varepsilon<|y|\leq 1\}} y\widetilde{N}(ds, dy)$$

where N denotes the Poisson measure associated to Lévy process, $\widetilde{N}(ds, dy) := N(ds, dy) - \nu(dy)ds$ is the compensated Poisson measure and the limit is a.s. and uniformly on compacts.

For the integers $i \geq 0$, we consider the following constants, provided they exist:

$$c_i := \sum_{k=i}^{\infty} \int_{\mathbb{R}} \frac{y^k}{k!} \nu(dy).$$

Observe that in particular

$$c_0 = \int_{\mathbb{R}} e^y \nu(dy),$$
$$c_1 = \int_{\mathbb{R}} (e^y - 1)\nu(dy),$$
$$c_2 = \int_{\mathbb{R}} (e^y - 1 - y)\nu(dy).$$

In order to $e^{-rt}e^{X_t}$ be a martingale, see for example [6], we must assume

$$\int_{|y|\geq 1} e^y \nu(dy) < \infty \text{ and } \gamma_0 = -\int_{\mathbb{R}} (e^y - 1 - y \mathbb{1}_{\{|y|<1\}})\nu(dy).$$

These conditions guarantee that ν has moments of all orders greater or equal than 2 and that we can write

$$L_t^0 = \int_0^t \int_{\mathbb{R}} y\widetilde{N}(ds, dy) - c_2 t.$$

So, in the following we will assume, without loosing generality, the model

$$X_t = x + (r - c_2)t - \frac{1}{2}\int_0^t \sigma_s^2 ds + \int_0^t \sigma_s (\rho dW_s + \sqrt{1-\rho^2} dB_s) + J_t$$

with

$$J_t := \int_0^t \int_{\mathbb{R}} y\widetilde{N}(ds, dy).$$

Recall that if $\int_{\mathbb{R}} |y|\nu(dy) = \infty$ we say that the process has infinite activity and infinite variation. In this case $c_1 := \int_{\mathbb{R}}(e^y - 1)\nu(dy)$ and $c_0 := \int_{\mathbb{R}} e^y \nu(dy)$ are infinite or not defined. If ν has first order moment, that is $\int_{\mathbb{R}} |y|\nu(dy) < \infty$, we say the model has infinite activity but finite variation and c_1 is finite. In this case we can consider $c_2 = c_1 - \int_{\mathbb{R}} y\nu(dy)$ and rewrite

$$\int_0^t \int_{\mathbb{R}} y\widetilde{N}(ds, dy) - c_2 t = \int_0^t \int_{\mathbb{R}} yN(ds, dy) - c_1 t.$$

Finally, if ν is finite, the model has finite activity and so, it is a Compound Poisson process with $\nu = \lambda Q$ for a certain probability measure Q and a certain constant $\lambda = \nu(\mathbb{R}) > 0$. In this case,

$$c_1 = \int_{\mathbb{R}} (e^y - 1)\nu(dy) = c_0 - \lambda$$

and

$$\int_0^t \int_{\mathbb{R}} yN(ds, dy) = \sum_{i=1}^{N_t} V_i,$$

where N is a $\lambda-$Poisson process and V_i are i.i.d. random variables with law Q, the law that produce the jumps.

Let $\mathscr{F}^J$ be the filtration generated by J. Note that this filtration is the same as the filtration generated by L^0 because the difference of these two processes is deterministic. Define now $\mathscr{F}$, the filtration associated to S, by

$$\mathscr{F} := \mathscr{F}^W \vee \mathscr{F}^B \vee \mathscr{F}^J.$$

We will consider our price model defined on the product of the canonical spaces of processes W, B and J. This means that

$$\omega := (\omega^W, \omega^B, \omega^J) \in \Omega := \Omega^W \times \Omega^B \times \Omega^J$$

and in the rest of the paper, any hypothesis on one of the spaces will mean that the property is true almost surely with respect to the other spaces.

Note that this is a very general stochastic volatility model because, being σ adapted to $\mathscr{F}^W \vee \mathscr{F}^J$, we are allowing jumps both in price and volatility. Recall the following facts:

- If we assume no jumps, that is $\nu = 0$, we have a generalization of correlated and non correlated stochastic volatility models in the sense that we do not assume a concrete dynamics for the volatility. This is the case treated in Sect. 2.
- If we restrict our model to the case σ adapted only to $\mathscr{F}^W$ we have a generalization of the Bates model in a double sense; on one hand we do not assume any concrete dynamics for the stochastic volatility and on other hand we are not assuming finite activity nor finite variation on ν.
- If we assume no correlation but presence of jumps we cover for example Heston-Kou model or any uncorrelated model with the addition of Lévy jumps in the price process with any Lévy measure ν.
- If $\sigma = 0$ but we have jumps, we cover the so called exponential Lévy models.

3.2 *Malliavin-Skorohod Type Calculus for Lévy Processes*

The literature on Malliavin calculus for Lévy processes is more recent and less extended. Here we follow closely [11] and [4]. A survey of this results can be found in [12]. We refer to these references for proofs of next results.

Let us denote $\mathbb{R}_0 := \mathbb{R} - \{0\}$. Consider the canonical version of the pure jump Lévy process J. It is defined on the space Ω^N given by the finite or infinite sequences of pairs $(t_i, x_i) \in (0, T] \times \mathbb{R}_0$ such that for every $\varepsilon > 0$ there is only a finite number of (t_i, x_i) with $|x_i| > \varepsilon$. Of course, t_i denotes a jump instant and x_i a jump size.

Consider $\omega^N \in \Omega^N$. Given $(t, x) \in [0, T] \times \mathbb{R}_0$ we can introduce a jump of size x at instant t to ω^N and call the new element

$$\omega^N_{t,x} := ((t, x), (t_1, x_1)(t_2, x_s), ...).$$

For a random variable $F \in L^2(\Omega^N)$, we define

$$T_{t,x} F(\omega^N) = F(\omega^N_{t,x})$$

and

$$D^N_{t,x} F = \frac{T_{t,x} F(\omega^N) - F(\omega^N)}{x}.$$

The operator D^N is closed and densely defined in $L^2(\Omega^N)$ and its domain $Dom\ D^N$ can be characterized by the fact that

$$F \in Dom\ D^N \Longleftrightarrow D^N F \in L^2(\Omega \times [0, T] \times \mathbb{R}_0, P \otimes ds \otimes x^2 \nu(dx)).$$

On other hand we define δ^N as the dual operator of D^N.

Given $u \in L^2(\Omega^W \times [0, T] \times \mathbb{R}, P \otimes ds \otimes x^2 \nu(dx))$, $\delta^N(u)$ is the element of $L^2(\Omega^N)$ characterized by

$$\mathbb{E}_N(F \delta^N(u)) = \mathbb{E}_N(\int_0^T \int_{\mathbb{R}} u_{t,x} D^N_{t,x} F x^2 \nu(dx) dt)$$

for any $F \in Dom\ D^N$. In particular $\mathbb{E}_N(\delta^N(u)) = 0$.

Let us denote $\delta^N_t(u) := \delta^N(u \mathbb{1}_{[0,t]})$.

As we have seen, D^N is an increment quotient operator and it is also known that δ^N_t is an extension of Itô integral in the sense that

$$\delta^N_t(u \mathbb{1}_{\mathbb{R}_0}) = \int_0^t \int_{\mathbb{R}} u(s, x) x \tilde{N}(ds, dx)$$

for predictable integrands u.

The following formulas will be helpful:

- If F, G and $F \cdot G$ belong to $Dom\ D^N$ we have

$$D^N(F \cdot G) = F D^N G + G D^N F + x D^N F D^N G.$$

- If $F \in Dom\, D^N$, $u \in Dom\, \delta^N$ and $u \cdot T_{t,x}F \in Dom\, \delta^N$ then

$$\delta^N(F \cdot u) = F\delta^N(u) - \int_0^T \int_{\mathbb{R}} u_{t,x} D^N_{t,x} F x^2 \nu(dx)dt - \delta^N(x \cdot u \cdot D^N F).$$

As in the Wiener case we define the space

$$\mathbb{L}^{1,2}_N := L^2([0, T] \times \mathbb{R}, Dom\, D^N),$$

that is the space of processes $u \in L^2([0, T] \times \mathbb{R} \times \Omega^N)$ such that $u_{t,x} \in Dom\, D^N$ for almost all (t, x) and $Du \in L^2(\Omega^N \times ([0, T] \times \mathbb{R})^2)$.

It can be proved that $\mathbb{L}^{1,2}_N \subseteq Dom\, \delta^N$ and

$$E_N(\delta^N(u)^2) \leq ||u||^2_{\mathbb{L}^{1,2}_N} := E_N(||u||^2_{L^2([0,T]\times\mathbb{R})}) + E_N(||D^N u||^2_{L^2(([0,T]\times\mathbb{R})^2)}).$$

Moreover we introduce the space $\mathbb{L}^{1,2}_{N,-}$ as the subspace of $\mathbb{L}^{1,2}_N$ of processes u such that the left-limits

$$u(s-, y) := \lim_{r\uparrow s, x\uparrow y} u(r, x)$$

and

$$D^{N,-}_{s,y} u(s-, y) := \lim_{r\uparrow s, x\uparrow y} D^N_{s,y} u(r, x)$$

exists $\mathbb{P}_N \otimes ds \otimes x^2\nu(dx)$—a.s. and belong to $L^2(\Omega^N \times [0, T] \times \mathbb{R})$.

Assume $u \in \mathbb{L}^{1,2}_{N,-}$ and $\int_0^T \int_{\mathbb{R}_0} |u(s-, y)||y|N(ds, dy) \in L^2(\Omega^N)$. Then, for any $t \in [0, T]$,

$$T^-_{s,y} u(s-, y) := u(s-, y) + y D^{N,-}_{s,y} u(s-, y) \in Dom\, \delta^N_t$$

and

$$\int_0^t \int_{\mathbb{R}} u(s-, y) y \tilde{N}(ds, dy) = \delta^N_t(T^-_{s,y} u(s-, y) \mathbb{1}_{\mathbb{R}_0}) + \int_0^t \int_{\mathbb{R}} D^{N,-}_{s,y} u(s-, y) y^2 \nu(dy)ds.$$

If u is predictable we have $D^{N,-}_{s,y} u(s-, y) = 0$. Hence, in this case,

$$\int_0^t \int_{\mathbb{R}} u(s-, y) y \tilde{N}(ds, dy) = \delta^N_t(u(s-, y) \mathbb{1}_{\mathbb{R}_0}).$$

3.3 The Hull and White Formula in the Lévy Case

Consider the following definitions in order to shorten the notation, for a suitable function F :

- $\Delta_x F(s, X_s, Y_s) := F(s, X_s + x, Y_s) - F(s, X_s, Y_s)$.
- $\Delta_{xx} F(s, X_s, Y_s) := F(s, X_s + x, Y_s) - F(s, X_s, Y_s) - x(\partial_x F)(s, X_s, Y_s)$.
- $\Delta F(s, X_s, Y_s) = F(s, X_s + x, Y_s) - F(s, X_s, Y_s) - (e^x - 1)(\partial_x F)(s, X_s, Y_s)$.

Then, we have the following decomposition of the price formula:

Theorem 2 *Assume*

- *(B1):* $\sigma^2 \in \mathbb{L}^{1,2}_{N,-} \cap \mathbb{L}^{1,2}_{W}$.
- *(B2):* $\sigma \in \mathbb{L}^{1,2}_{W}$.
- *(B3): For any* $t \in [0, T]$, $\int_t^T \mathbb{E}_t((\int_t^s \sigma_u^2 du)^{-2}) ds < \infty$.

Then we have

$$\begin{aligned} V_t &= \mathbb{E}_t(BS(t, X_t, v_t)) \\ &+ \frac{\rho}{2} \mathbb{E}_t(\int_t^T e^{-r(s-t)} (\partial_x^3 - \partial_x^2) BS(s, X_s, \bar{\sigma}_s) \Lambda_s ds) \\ &+ \mathbb{E}_t(\int_t^T \int_{\mathbb{R}} e^{-r(s-t)} \Delta BS(s, X_{s-}, \bar{\sigma}_s) \nu(dy) ds) \\ &+ \mathbb{E}_t(\int_t^T \int_{\mathbb{R}} e^{-r(s-t)} D^{N,-}_{s,y} \Delta_y BS(s, X_{s-}, \bar{\sigma}_s) y \nu(dy) ds). \end{aligned}$$

Remark 2 We can consider the following particular cases:

1. Observe that we cannot split the third term in two terms because in the general case

$$\mathbb{E}_t(\int_t^T \int_{\mathbb{R}} e^{-r(s-t)} \Delta_y BS(s, X_{s^-}, \bar{\sigma}_s) \nu(dy) ds)$$

and

$$\mathbb{E}_t(\int_t^T \int_{\mathbb{R}} e^{-r(s-t)} (e^y - 1) \partial_x BS(s, X_{s^-}, \bar{\sigma}_s) \nu(dy) ds)$$

are not necessarily convergent.

2. Observe that if in the previous theorem we assume $\int_{\mathbb{R}} |y| \nu(dy) < \infty$, that is, finite variation, we obtain

$$
\begin{aligned}
V_t &= \mathbb{E}_t(BS(t, X_t, v_t)) \\
&+ \frac{\rho}{2}\mathbb{E}_t(\int_t^T e^{-r(s-t)}(\partial_x^3 - \partial_x^2)BS(s, X_s, \bar{\sigma}_s)\Lambda_s ds) \\
&- \mathbb{E}_t(\int_t^T \int_{\mathbb{R}} e^{-r(s-t)}(e^y - 1)\partial_x BS(s, X_s, \bar{\sigma}_s)\nu(dy)ds) \\
&+ \mathbb{E}_t(\int_t^T \int_{\mathbb{R}} e^{-r(s-t)} T^-_{s,y}\Delta_y BS(s, X_{s^-}, \bar{\sigma}_s)\nu(dy)ds),
\end{aligned}
$$

that is exactly the formula obtained in [5] for the finite activity case, that in fact is valid in the finite variation case.

3. If the volatility process is independent from price jumps, we have

$$
D^{N,-}_{s,y}\Delta_y BS(s, X_{s-}, \bar{\sigma}_s) = 0
$$

and we obtain

$$
\begin{aligned}
V_t &= \mathbb{E}_t(BS(t, X_t, \bar{\sigma}_t)) \\
&+ \frac{\rho}{2}\mathbb{E}_t(\int_t^T e^{-r(s-t)}(\partial_x^3 - \partial_x^2)BS(s, X_s, \bar{\sigma}_s)\Lambda_s ds) \\
&+ \mathbb{E}_t \int_t^T \int_{\mathbb{R}} e^{-r(s-t)}\Delta BS(s, X_{s^-}, \bar{\sigma}_s)\nu(dy)ds,
\end{aligned}
$$

that generalizes the formula in [3]. As in the previous remark, only in the finite variation case we recuperate exactly the formula in [3]. This formula covers Bates model and any correlated model with any type of Lévy jumps in the price process.

4. If moreover, the volatility process is independent from the price process, that is, $\rho = 0$, we obtain

$$
V_t = \mathbb{E}_t(BS(t, X_t, v_t)) + \mathbb{E}_t \int_t^T \int_{\mathbb{R}} e^{-r(s-t)}\Delta BS(s, X_{s^-}, \bar{\sigma}_s)\nu(dy)ds.
$$

This covers all the so called uncorrelated models plus jumps (Heston-Kou model for example) and in the particular case of constant volatility, the so called exponential Lévy models. In the jump part we can consider infinite activity jumps as CGMY model (for $Y \geq 0$) or Meixner model for example.

Proof We follow the same general idea of Theorem 1. The necessary ad-hoc Itô formula, see [9], is now

$$
\begin{aligned}
F(t,X_t,Y_t) &= F(0,X_0,Y_0) + \int_0^t \partial_s F(s,X_s,Y_s)ds + \delta_t^{W,B}(\partial_x F(\cdot,X_\cdot,Y_\cdot)\sigma_\cdot) \\
&+ \int_0^t \partial_x F(s,X_s,Y_s)(r - c_2 - \frac{\sigma_s^2}{2})ds - \int_0^t \partial_y F(s,X_s,Y_s)\sigma_s^2 ds \\
&+\rho \int_0^t \partial_{xy} F(s,X_s,Y_s)\Lambda_s ds + \frac{1}{2}\int_0^t \partial_x^2 F(s,X_s,Y_s)\sigma_s^2 ds \\
&+ \int_0^t \int_{\mathbb{R}_0} (\Delta_y F(s,X_{s-},Y_s) - y(\partial_y F)(s,X_{s-},Y_s))\nu(dy)ds \\
&+\delta_t^N(\frac{\Delta_y F(s,X_{s-},Y_s)}{y} \mathbb{1}_{\mathbb{R}_0}(y)) + \delta_t^N(D_{s,y}^{N,-}\Delta_y(s,X_{s-},Y_s)) \\
&+ \int_0^t \int_{\mathbb{R}} D_{s,y}^{N,-}\Delta_y F(s,X_{s-},Y_s) y\nu(dy)ds.
\end{aligned}
$$

To prove it, fix first $\varepsilon > 0$, and consider the process

$$
\begin{aligned}
X_t^\varepsilon &:= x + (r - c_2)t - \frac{1}{2}\int_0^t \sigma_s^2 ds + \int_0^t \sigma_s(\rho dW_s + \sqrt{1-\rho^2}dB_s) \\
&+ \int_0^t \int_{|x|>\varepsilon} x\tilde{N}(ds,dx).
\end{aligned}
$$

This process has a finite number of jumps and converges a.s. and in L^2 sense to X_t.

Denote by T_i^ε the jump instants, and write $T_0^\varepsilon := 0$. Then

$$
\begin{aligned}
F(T_{i+1}^\varepsilon, X_{T_{i+1}^\varepsilon}^\varepsilon, Y_{T_{i+1}^\varepsilon}) - F(T_i^\varepsilon, X_{T_i^\varepsilon}^\varepsilon, Y_{T_i^\varepsilon}) &= \int_{T_i^\varepsilon}^{T_{i+1}^\varepsilon -} dF(s, X_s^\varepsilon, Y_s) \\
+F(T_{i+1}^\varepsilon, X_{T_{i+1}^\varepsilon}^\varepsilon, Y_{T_{i+1}^\varepsilon}) - F(T_{i+1}^\varepsilon, X_{T_{i+1}^\varepsilon -}^\varepsilon, Y_{T_{i+1}^\varepsilon}).
\end{aligned}
$$

On the stochastic interval $[T_j^\varepsilon, T_{j+1}^\varepsilon[$ we can apply the anticipative Itô formula for continuous process presented in Sect. 2. Then we have that

$$
\partial_x F(s, X_{s^-}, Y_s)\sigma_s \mathbb{1}_{[0,t]}(s) \in Dom\, \delta^{W,B}
$$

and

$$
\begin{aligned}
F(t,X_t^\varepsilon,Y_t) &= F(0,X_0,Y_0) + \int_0^t \partial_s F(s,X_s^\varepsilon,Y_s)ds \\
&+ \int_0^t \partial_x F(s,X_s^\varepsilon,Y_s)(r - \frac{\sigma_s^2}{2} - c_2)ds + \delta_t^{W,B}\left(\partial_x F(s,X_{s^-}^\varepsilon,Y_s)\sigma_s\right) \\
&- \int_0^t \int_{\{|x|>\varepsilon\}} \partial_x F(s,X_s^\varepsilon,Y_s)x\nu(dx)ds - \int_0^t \partial_y F(s,X_s^\varepsilon,Y_s)\sigma_s^2 ds
\end{aligned}
$$

$$+\rho\int_0^t \partial_{xy}F(s,X_s^\varepsilon,Y_s)\Lambda_s ds+\frac{1}{2}\int_0^t \partial_x^2 F(s,X_s^\varepsilon,Y_s)\sigma_s^2 ds$$
$$+\sum_i [F(T_i^\varepsilon,X_{T_i^\varepsilon}^\varepsilon,Y_{T_i^\varepsilon})-F(T_i^\varepsilon,X_{T_i^\varepsilon-}^\varepsilon,Y_{T_i^\varepsilon})].$$

Of course we can write

$$\sum_i [F(T_i^\varepsilon,X_{T_i^\varepsilon}^\varepsilon,Y_{T_i^\varepsilon})-F(T_i^\varepsilon,X_{T_i^\varepsilon-}^\varepsilon,Y_{T_i^\varepsilon})]=\int_0^t\int_{|x|>\varepsilon}\Delta_x F(s,X_{s-},Y_s)N(ds,dx).$$

Then,

$$\sum_i [F(T_i^\varepsilon,X_{T_i^\varepsilon}^\varepsilon,Y_{T_i^\varepsilon})-F(T_i^\varepsilon,X_{T_i^\varepsilon-}^\varepsilon,Y_{T_i^\varepsilon})]-\int_0^t\int_{|x|>\varepsilon}\partial_x F(s,X_{s-}^\varepsilon,Y_s)x\nu(dx)ds$$

$$=\int_0^t\int_{|x|>\varepsilon}\Delta_x F(s,X_{s-}^\varepsilon,Y_s)\tilde{N}(ds,dx)+\int_0^t\int_{|x|>\varepsilon}\Delta_{xx} F(s,X_{s-}^\varepsilon,Y_s)\nu(dx)ds.$$

Observe that this equality is the crucial step of the proof. Only introducing $\Delta_{xx}F(s,X_{s-}^\varepsilon,Y_s)$ we become able to apply succesfully the dominated convergence theorem, even if Y has no jumps.

Using the relation between δ^N and the integral with respect to $\tilde{N}$ we have

$$\int_0^t\int_{|x|>\varepsilon}\Delta_x F(s,X_{s-}^\varepsilon,Y_s)\tilde{N}(ds,dx)$$
$$=\delta_t^N(T_{s,x}^-\frac{\Delta_x F(s,X_{s-}^\varepsilon,Y_s)}{x}\mathbb{1}_{\{|x|>\varepsilon\}})$$
$$+\int_0^t\int_{|x|>\varepsilon}D_{s,x}^{N,-}\frac{\Delta_x F(s,X_{s-}^\varepsilon,Y_s)}{x}x^2\nu(dx)ds.$$

And using mean value theorem and the fact that first and second derivatives of F are bounded we have

$$|T_{s,x}^-\frac{\Delta_x F(s,X_{s-}^\varepsilon,Y_s)}{x}|=|\frac{\Delta_x F(s,X_{s-}^\varepsilon,T_{s,x}^-Y_s)}{x}|\le C,$$

$$|D_{r,y}^{N,-}\frac{\Delta_x F(s,X_{s-}^\varepsilon,T_{s,x}^-Y_s)}{x}|\le C|D_{r,y}^{N,-}T_{s,x}^-Y_s|=C\int_s^T|D_{r,y}^{N,-}T_{s,x}^-\sigma_u^2|du$$

and

$$|D_{s,x}^{N,-}\frac{\Delta_x F(s,X_{s-}^\varepsilon,Y_s)}{x}|\le C|D_{s,x}^{N,-}Y_s|=C\int_s^T|D_{s,x}^{N,-}\sigma_r^2|dr,$$

for a generic constant C.

Finally, using (B1) and the dominated convergence theorem the right hand side of the previous equality converges when ε goes to 0. The other terms converge also by the dominated convergence theorem, and the Itô formula follows.

Then, following the same steps of the proof of Theorem 1, after applying this last ad-hoc Itô formula, taking conditional expectations, using the fact that Skorohod type integrals have zero expectation and multiplying by e^{rt} we obtain

$$
\begin{aligned}
&\mathbb{E}_t(e^{-r(T-t)}BS(T, X_T, \bar{\sigma}_T^\delta)\phi(\frac{X_T}{n})) \qquad (1)\\
&= \mathbb{E}_t(BS(t, X_t, \bar{\sigma}_t^\delta)\phi(\frac{X_t}{n}))\\
&+\mathbb{E}_t(\int_t^T e^{-r(s-t)}A_n(s)ds)\\
&+\frac{\rho}{2}\mathbb{E}_t(\int_t^T e^{-r(s-t)}(\partial_x^3 - \partial_x^2)BS(s, X_s, \bar{\sigma}_s^\delta)\phi_n(\frac{X_s}{n})\Lambda_s ds)\\
&+\frac{\rho}{2}\mathbb{E}_t(\int_t^T e^{-r(s-t)}(\partial_x^2 - \partial_x)BS(s, X_s, \bar{\sigma}_s^\delta)\frac{1}{n}\phi'(\frac{X_s}{n})\Lambda_s ds)\\
&-c_2\mathbb{E}_t(\int_t^T e^{-r(s-t)}\partial_x BS(s, X_s, \bar{\sigma}_s^\delta)\phi(\frac{X_s}{n})ds)\\
&+\mathbb{E}_t(\int_t^T\int_{\mathbb{R}} e^{-r(s-t)}\Delta_{yy}BS(s, X_{s^-}, \bar{\sigma}_s^\delta)\phi(\frac{X_{s-}}{n})\nu(dy)ds)\\
&+\mathbb{E}_t(\int_t^T\int_{\mathbb{R}} e^{-r(s-t)}D_{s,y}^{N,-}\frac{\Delta_y BS(s, X_{s^-}, \bar{\sigma}_s^\delta)\phi(\frac{X_{s-}}{n})}{y}y^2\nu(dy)ds).
\end{aligned}
$$

And as in Theorem 1, applying the dominated convergence theorem, letting first $n \uparrow \infty$ and then $\delta \downarrow 0$ we obtain the result.

To assure the dominated convergence, we have to treat the last three terms of (1) as a unique term and separate it in two integrals, one on $|y| \le 1$ and the other on $|y| > 1$.

In the case $|y| > 1$, things simplify and we obtain

$$
\begin{aligned}
&-\mathbb{E}_t(\int_t^T\int_{|y|>1} e^{-r(s-t)}\partial_x BS(s, X_s, \bar{\sigma}_s^\delta)\phi(\frac{X_s}{n})(e^y - 1)\nu(dy)ds)\\
&+\mathbb{E}_t(\int_t^T\int_{|y|>1} e^{-r(s-t)}T_{s,y}^{N,-}\Delta_y BS(s, X_{s^-}, \bar{\sigma}_s^\delta)\phi(\frac{X_{s-}}{n})\nu(dy)ds).
\end{aligned}
$$

For the first term we use that $\partial_x BS(s, X_s, \bar{\sigma}_s^\delta)$ is bounded by e^{X_s} and for the second term we use the fact that

$$|T_{s,y}^{N,-}\Delta_y BS(s, X_{s^-}, \bar{\sigma}_s^\delta)| \le 2K + e^{X_{s-}+y} + e^{X_{s-}}$$

and that

$$\int_{\{|y|>1\}} e^{y}\nu(dy) < \infty.$$

In the case $|y| \le 1$, the fifth term in the right hand side of (1) is bounded because $\partial_x BS$ is bounded. The sixth term can be written as

$$\frac{1}{2}\mathbb{E}_t\Big(\int_t^T \int_{|y|\le 1} e^{-r(s-t)}\partial_x^2 BS(s, X_s+\alpha, \bar{\sigma}_s^{\delta})\phi(\frac{X_{s-}}{n})y^2\nu(dy)ds\Big)$$
$$= \frac{1}{2}\mathbb{E}_t\Big(\int_t^T \int_{|y|\le 1} e^{-r(s-t)}(\partial_x^2 - \partial_x)BS(s, X_s+\alpha, \bar{\sigma}_s^{\delta})\phi(\frac{X_{s-}}{n})y^2\nu(dy)ds\Big)$$
$$+\frac{1}{2}\mathbb{E}_t\Big(\int_t^T \int_{|y|\le 1} e^{-r(s-t)}\partial_x BS(s, X_s+\alpha, \bar{\sigma}_s^{\delta})\phi(\frac{X_{s-}}{n})y^2\nu(dy)ds\Big)$$

where $|\alpha| \le |y|$.

The second term on the right hand side of this last expression is bounded because $\partial_x BS$ is bounded. For the first term we use Lemma 2 in [3] as in Theorem 1 and we bound it by

$$C\mathbb{E}_t\Big(\int_t^T \int_{|y|\le 1} \sqrt{\frac{1}{Y_t}}y^2\nu(dy)ds\Big),$$

for a certain constant C. Hypothesis (B3) guarantees the convergence of this integral, because

$$\mathbb{E}_t\Big(\int_t^T \int_{|y|\le 1} \sqrt{\frac{1}{Y_t}}y^2\nu(dy)ds\Big) = C(T-t)\mathbb{E}_t\Big(\sqrt{\frac{1}{Y_t}}\Big)$$

and

$$\mathbb{E}_t\Big(\sqrt{\frac{1}{Y_t}}\Big) \le \Big(\mathbb{E}_t\big(\frac{1}{Y_t^2}\big)\Big)^{\frac{1}{4}} \le \Big(\mathbb{E}_t\big((\int_t^s \sigma_u^2 du)^{-2}\big)\Big)^{\frac{1}{4}}$$

and so, the term is bounded by (B3).

Finally, the last term of (1) can be bounded by

$$\mathbb{E}_t\Big(\int_t^T \int_{|y|\le 1} |D_{s,y}^{N,-}(\partial_x BS)(s, X_{s^-}+\alpha, \bar{\sigma}_s^{\delta})|y^2\nu(dy)ds\Big)$$
$$= \mathbb{E}_t\Big(\int_t^T \int_{|y|\le 1} |D_{s,y}^{N,-}(\partial_x BS)(s, X_{s^-}+\alpha, \sqrt{\frac{Y_s+\delta}{T-s}})|y^2\nu(dy)ds\Big)$$
$$= \mathbb{E}_t\Big(\int_t^T \int_{|y|\le 1} |(\partial_{x\sigma} BS)(s, X_{s^-}+\alpha, \sqrt{\frac{\theta_{s,y}+\delta}{T-s}})|\frac{|D_{s,y}^{N,-}Y_s|}{2(T-s)\sqrt{\frac{\theta_{s,y}+\delta}{T-s}}}y^2\nu(dy)ds\Big)$$

$$= \frac{1}{2}\mathbb{E}_t\Big(\int_t^T \int_{|y|\le 1} |(\partial_x(\partial_x^2 - \partial_x)BS)(s, X_{s^-} + \alpha, \sqrt{\frac{\theta_{s,y}+\delta}{T-s}})||D_{s,y}^{N,-} Y_s|y^2\nu(dy)ds\Big), \tag{2}$$

where $|\alpha| \le |y|$ and $\theta_{s,y}$ is a quantity between Y_s and $T_{s,y}^{N,-} Y_s$.

Now, we cannot apply directly Lemma 2 in [3], but mimicking the proof we have that the last integral is less or equal than

$$C\mathbb{E}_t\Big(\int_t^T \int_{|y|\le 1} \frac{|D_{s,y}^{N,-} Y_s|}{\int_t^s \sigma_u^2 du} y^2\nu(dy)ds\Big).$$

Then, applying Cauchy-Schwarz inequality, this expression is bounded by

$$C\Big(\mathbb{E}_t\Big(\int_t^T \int_{|y|\le 1} \frac{1}{(\int_t^s \sigma_u^2 du)^2} y^2\nu(dy)ds\Big)\Big)^{\frac{1}{2}}\Big(\mathbb{E}_t\Big(\int_t^T \int_{|y|\le 1} |D_{s,y}^{N,-} Y_s|^2 y^2\nu(dy)ds\Big)\Big)^{\frac{1}{2}}.$$

The first term of this product is bounded by (B3) and the second one by (B1)

Remark 3

1. As in the case of Theorem 1, (B2) can be changed by (A2′).
2. If σ not depends on jumps, (B3) reduces to $\mathbb{E}_t((\int_t^T \sigma_u^2 du)^{-\frac{1}{2}}) < \infty$, that it is weaker than (A2′).
3. In the case of finite variation, (B3) is not necessary.
4. In the complete general case, but only in this case, (B1) and (A2′) are not enough. An alternative treatment of (2), using (A2′), is to bound directly

$$|(\partial_x(\partial_x^2 - \partial_x)BS)(s, X_{s^-} + \alpha, \sqrt{\frac{\theta_{s,y}+\delta}{T-s}})| \le e^{X_{s^-}+\alpha}\Big(\frac{1}{\sqrt{Y_s}} + \frac{1}{Y_s}\Big).$$

   So, we can decompose this term in two new terms. The term with $Y_s^{-\frac{1}{2}}$ can be treated easily and it is bounded with no other requirements than (B1) and (A2′). But the term with Y_s^{-1} requires to assume, alternatively to (B3), the following hypothesis,

$$\mathbb{E}_t\Big(\int_t^T \int_{|y|\le 1} \frac{|D_{s,y}^{N,-} Y_s|}{Y_s} y^2\nu(dy)ds\Big) < \infty,$$

   that using (A2′) is equivalent to assume

$$(B4) : \mathbb{E}_t\Big(\int_t^T \int_{|y|\le 1} \frac{|D_{s,y}^{N,-} Y_s|}{T-s} y^2\nu(dy)ds\Big) < \infty.$$

   Note that this last hypothesis is stronger than (B1). So, we need (B1), (A2′) and (B4) to guarantee the complete general case.

Acknowledgments This paper was partially written during a four month stage in Center for Advanced Study (CAS) at the Norwegian Academy of Science and Letters. I thank CAS for the support and the kind hospitality.

References

1. Alòs, E.: A generalization of the Hull and White formula with applications to option pricing approximation. Finance Stochast. **10**(3), 353–365 (2006)
2. Alòs, E., Nualart, D.: An extension of Itô formula for anticipating processes. J. Theor. Probab. **11**(2), 493–514 (1998)
3. Alòs, E., León, J.A., Vives, J.: On the short-time behavior of the implied volatility for jump diffusion models with stochastic volatility. Finance Stochast. **11**(4), 571–589 (2007)
4. Alòs, E., León, J., Vives, J.: An anticipating Itô formula for Lévy processes. Lat. Am. J. Probab. Math. Stat. **4**, 285–305 (2008)
5. Alòs, E., León, J.A., Pontier, M., Vives, J.: A Hull and White formula for a general stochastic volatility jump diffusion model with applications to the study of the short time behavior of the implied volatility. J. Appl. Math. Stoch. Anal. ID 359142, 17 (2008)
6. Cont, R., Tankov, P.: Financial modelling with jump processes. Chapman-Hall/CRC (2004)
7. Fouque, J.P., Papanicolaou, G., Sircar, R.: Derivatives in financial markets with stochastic volatility. Cambridge (2000)
8. Gulisashvili, A.: Analytically Tractable Stochastic Stock Price Models. Springer (2012)
9. Jafari, H., Vives, J.: A Hull and White formula for a stochastic volatility Lévy model with infinite activity. Commun. Stochast. Anal. **7**(2), 321–336 (2013)
10. Nualart, D.: The Malliavin Calculus and Related Topics. Springer (2006)
11. Solé, J.L., Utzet, F., Vives, J.: Canonical Lévy process and Malliavin calculus. Stochast. Processes Appl. **117**(2), 165–187 (2007)
12. Vives, J.: Malliavin calculus for Lévy processes: a survey. Proceedings of the 8th Conference of the ISAAC-2011. Rendiconti del Seminario Matematico di Torino, 71 (2013)

Part II
Insurance

Statistical Estimation Techniques in Life and Disability Insurance—A Short Overview

Boualem Djehiche

Abstract This is a short introduction to some basic aspects of statistical estimation techniques known as graduation technique in life and disability insurance.

Keywords Life insurance · Disability insurance · Claims reserving · Mortality modeling · Thiele's equation

Mathematics Subject Classification 2010 90B30 · 60J75 · 62M10

1 Life Insurance

By life insurance policy or contract we mean any form of person insurance contract over a (long) period of time such as life or pension and disability or sickness coverage. In such products, premiums and benefits are typically contingent upon transitions of the policyholder between a number of states stated in the contract. Thereof the use of the powerful (semi)-Markov chain theory to carry out the valuation of insurance contracts and estimation of the underlying rates. We first give a short introduction to the basic constituents of a life insurance contract and related reserving. Then we single out the main parameters that control the evolution of the life insurance contract and focus on their statistical estimation. These parameters are the mortality rate and disability inception and recovery rates. Due to lack of space, the reader is referred to the list of references for an update of recent developments in claims reserving techniques for life and disability insurance. A detailed account for basic life insurance contracts can be found in the papers [11–15] by Norberg. A very short summary is displayed in Sects. 1.1–1.6, below.

B. Djehiche (✉)
Department of Mathematics, KTH Royal Institute of Technology,
SE-100 44, Stockholm, Sweden
e-mail: boualem@math.kth.se

M. Eddahbi et al. (eds.), *Statistical Methods and Applications in Insurance and Finance*, Springer Proceedings in Mathematics & Statistics 158,
DOI 10.1007/978-3-319-30417-5_5

1.1 A Markov Chain Model of a Life Insurance Contract

Let $E = \{0, 1, 2, \ldots, m\}$ be the (finite) set of possible states of the policy. Starting at 0, the policy is assumed to be in one and only one state at each time. Let $X(t)$ denote the state of the policy at time $t \in [0, n]$. We assume that the process X is right-continuous with a finite number of jumps, with transition probability

$$p_{ij}(s,t) = P[X(t) = i | X(s) = j], \quad i, j \in E, \quad 0 \le s \le t \le n, \tag{1}$$

and transition intensity

$$\mu_{ij}(t) := \lim_{h \downarrow 0} \frac{p_{ij}(t, t+h)}{h}, \quad i \neq j. \tag{2}$$

The total transition intensity from state i at time t is

$$\mu_{i\cdot}(t) = \sum_{k:k \neq i} \mu_{ik}(t) \tag{3}$$

so that

$$p_{ii}(t, t+dt) = 1 - \mu_{i\cdot}(t)dt + o(t).$$

1.1.1 Basic Kolmogorov Equations

The transition probabilities $(p_{ij}(s,t),\ i, j \in E,\ 0 \le s \le t \le n)$ satisfy the following equations.

A. The Kolmogorov backward equation: for $s \le t$,

$$\begin{cases} \frac{\partial p_{ij}}{\partial s}(s,t) = \mu_{i\cdot}(s) p_{ij}(s,t) - \sum_{k:k\neq i} \mu_{ik}(s) p_{kj}(s,t), \\ p_{ij}(t,t) = \delta_{ij}. \end{cases} \tag{4}$$

B. The Kolmogorov forward equation: for $s \le t$,

$$\begin{cases} \frac{\partial p_{ij}}{\partial t}(s,t) = -p_{ij}(s,t)\mu_{j\cdot}(t) + \sum_{k:k\neq i} p_{ik}(s,t)\mu_{kj}(t), \\ p_{ij}(s,s) = \delta_{ij}. \end{cases} \tag{5}$$

C. The Chapman-Kolmogorov equation

$$p_{ik}(s,u) = \sum_{j \in E} p_{ij}(s,t) p_{jk}(t,u), \quad s \le t \le u. \tag{6}$$

The key parameter in this Markov chain framework is the *transition intensity* which is the object of our statistical inference study.

1.2 Examples

1.2.1 Single Life with One Cause of Death (One Absorbing State)

In this model $E = \{0, 1\}$, where state 0 = alive, state 1= dead (absorbing state). If T denotes the life length of a person with survival probability

$$\bar{F}(t) = P(T > t),$$

the Markov chain counts the number of deaths:

$$X(t) = \mathbb{1}_{\{T \leq t\}}, \quad t \in [0, n],$$

with transition probability

$$p_{00}(s, t) = \frac{\bar{F}(t)}{\bar{F}(s)} = e^{-\int_s^t \mu(u)du}.$$

μ is called *mortality intensity (rate or force)*. Its estimation from data is of central importance in life insurance Fig. 1

1.2.2 Single Life with m Causes of Death (m Absorbing States)

In this model $E = \{0, 1, \ldots, m\}$, where state 0 = alive, state j= dead with cause j (absorbing state). These absorbing states model different causes of death such as death by "car accident", "normal death" or "death caused by a disease" etc. Fig. 2.

The total mortality intensity is

$$\mu_{0\cdot}(t) := \mu(t) = \sum_{j=1}^{r} \mu_j(t), \qquad (7)$$

where, $\mu_j(t) := \mu_{0j}(t)$ denotes the mortality rate for death with cause j. This is nothing but the transition intensity from state 0 (alive) to the absorbing state j.

Fig. 1 Single life with one cause of death

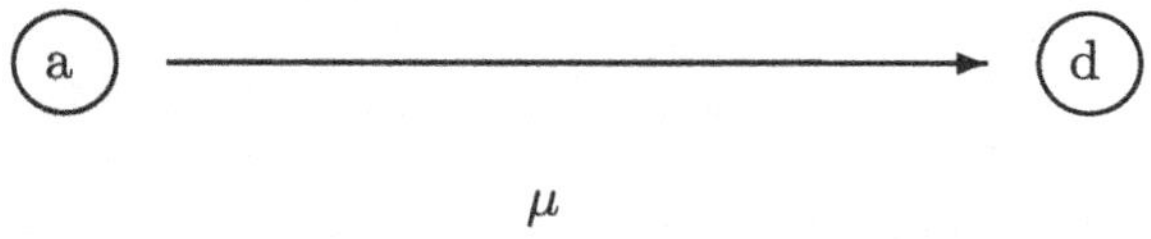

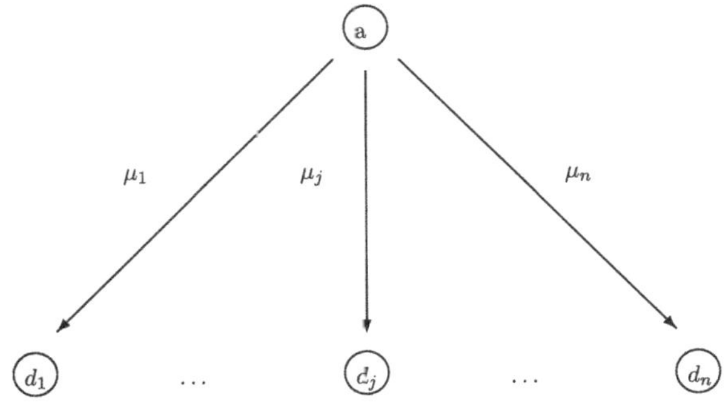

Fig. 2 Single life with m causes of death

The probability that an s years old person will die from cause j before age t is then

$$p_{0j}(s,t) = \int_s^t e^{-\int_s^u \mu(\tau)d\tau} \mu_j(u)du. \tag{8}$$

1.2.3 Disability, Recovery and Death

This model is widely used to analyze insurance contracts with payments depending on the state of the health of the insured. For example

- Sickness insurance that provides an annuity benefit during disability periods.
- Life insurance with premium waiver during disability.
- Pension with additional benefits to other members of the family.

The possible states are a = alive/active, i=invalid/unemployed, and d=dead/recovered or any other suitable labeling Fig. 3.

1.3 Payment Streams and Reserving Techniques

Let X be the Markov chain with intensities μ_{ij} associated with an insurance contract. Let

$$I_j(t) = \mathbb{1}_{\{X(t)=j\}}, \quad t \in [0, T],$$

denote the indicator process of whether the policy is in state j or not, and

$$N_{ij}(t) = \#\{s\colon X(s^-) = i,\ X(s) = j,\ s \in (0, t]\}, \quad i \neq j,$$

denote the number of transitions from state i to state j during the time interval $(0, t]$.

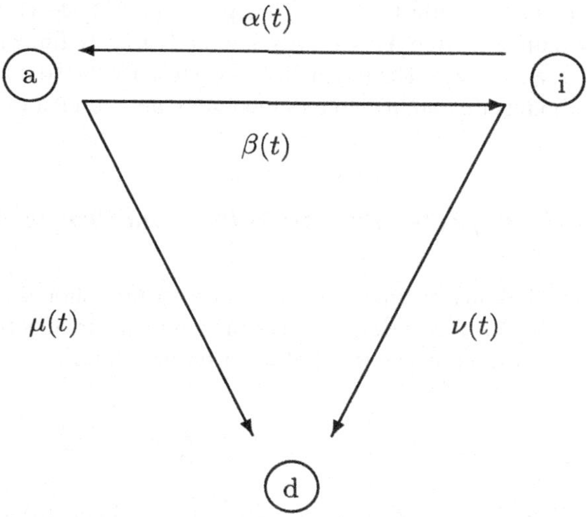

Fig. 3 Three possible states of a life insurance contract

We have

$$dI_j(t) = dN_{\cdot j}(t) - dN_{j\cdot}(t), \tag{9}$$

where,

$$N_{\cdot j}(t) := \sum_{k;k\neq j} N_{kj}(t), \qquad N_{j\cdot}(t) := \sum_{k;k\neq j} N_{jk}(t).$$

We have, for $t \leq u$

$$\begin{aligned} E[I_j(u)|X(t) = i] &= p_{ij}(t,u), \\ E[dN_{jk}(u)|X(t) = i] &= p_{ij}(t,u)\mu_{jk}(u)du. \end{aligned} \tag{10}$$

A standard payment stream A (benefits less premiums) has usually the following form:

$$dA(t) := \sum_j \left(I_j(t)dA_j(t) + \sum_{k;k\neq j} a_{jk}(t)dN_{jk}(t) \right), \tag{11}$$

where,

$$dA_j(t) := a_j(t)dt + (A_j(t) - A_j(t^-)) = a_j(t)dt + \Delta A_j(t) \tag{12}$$

specifies the so-called *general life annuity payment* i.e. payments due during sojourn in state j. The payment $a_j(t)$ is the rate of a state-wise annuity payable continuously at time t, while the lump sum payment $\Delta A_j(t)$ is an endowment at time t. The annuity function A_j is usually assumed to have a finite number of discontinuity points $\{t_1, t_2, \ldots, t_q\}$. The payments $a_{jk}(t)$ specify the so-called *general life assurance* i.e. amounts that are payable immediately upon transition from state j to state k.

1.4 Expected Present Values and Prospective Reserves

The liability at time t for which the insurer should provide a reserve (prospective reserve) is the present value of the payment streams (future benefits less premiums) A over the lifespan $[t, n]$ of the insurance contract:

$$V(t) = \int_t^n e^{-\int_t^s r(u)du} dA(s). \tag{13}$$

When the policy is in state i at time t, then, in view of Eq. (10), the state-wise *prospective reserve* is

$$\begin{aligned} V_i(t) &:= E[V(t)|X(t) = i] = \textstyle\int_t^n e^{-\int_t^s r(u)du} E[dA(s)|X(t) = i] \\ &= \textstyle\int_t^n e^{-\int_t^s r(u)du} \sum_j p_{ij}(t, s) \left(dA_j(t) + \sum_{k;k\neq j} a_{jk}(s)\mu_{jk}(s)ds\right), \end{aligned} \tag{14}$$

when r, a_j, a_{ik} are all deterministic function.
Written in differential form, V_j satisfies the following Feynman-Kac type formula known as **the backward Thiele's differential equation**:

$$\begin{cases} \frac{dV_i}{dt}(t) = (r(t) + \mu_{i\cdot}(t))V_i(t) - \sum_{j;j\neq i} \mu_{ij}(t)V_j(t) - a_i(t) - \sum_{j;j\neq i} a_{ij}(t)\mu_{ij}(t), \\ \qquad t \in (t_{p-1}, t_p), \quad p = 1, \ldots, q, \\ \Delta V_j(t_p) = -\Delta A_j(t_p), \quad p = 1, 2, \ldots, q, \quad i \in E, \\ V_j(n) = 0. \end{cases} \tag{15}$$

This equation admits an explicit solution only for a few uninteresting/trivial insurance contracts. In most cases it is solved using a numerical integration recipe. A fourth order "Runge-Kutta" procedure seems to work efficiently in almost all practical situations.

Thiele's equation can be recast in the following form "preferred by actuaries"

$$-a_i(t)dt = dV_i(t) - r(t)V_i(t)dt + \sum_{j;j\neq i} R_{ij}(t)\mu_{ij}(t)dt \tag{16}$$

where,

$$R_{ij}(t) = a_{ij}(t) + V_j(t) - V_i(t), \tag{17}$$

is the so-called "Sum-at-Risk" associated with a possible transition from state i to state j.

- The term $\sum_{j;j\neq i} R_{ij}(t)\mu_{ij}(t)dt$ is called the "risk premium" in $(t, t+dt)$.
- The term $dV_i(t) - r(t)V_i(t)dt$ is called the "savings premium" in $(t, t+dt)$.

1.5 The Equivalence Principle (aka Fairness Constraint)

The equivalence principle of insurance states that the expected present values of premiums and benefits should be equal. That is, roughly speaking, premiums and benefits should balance on the average. In our context this principle states that

$$V_0(0) = -A_0(0). \tag{18}$$

This condition imposes a constraint on the contractual payments a_j, A_j and a_{ij} to design a premium level for given benefits. Noting that $A_0(0^-) = 0$, we easily see that Eq. (18) is equivalent to

$$V_0(0^-) := E\left[\int_{0^-}^{n} e^{-\int_0^s r(u)du} dA(s)\right] = 0. \tag{19}$$

The state-wise prospective reserve $V(t)$ can be seen as the value function of a *singular* control problem subject to the fairness constraint, where the control parameter is the process $A(t)$.

1.6 First and Second Order Reserving Bases

The jump intensities μ_{ij} (purely actuarial parameters or liability driving parameter) and the discounting rate r which reflects the "expected return" of the investment portfolio (the main driver of the asset side) constitute the so-called *reserving basis*:

- **First order technical basis (prudent or conservative)**. This is a set of assumptions about the portfolio return (or just an interest rate that reflects the market value of the cash flow), r, the transition rates μ_{ij} (including mortality rates), costs and other relevant technical parameters etc. These assumptions are meant to yield premiums and reserves that include a high safety loading that hedges against worst case scenarios. The first order premiums and reserves are usually higher than experience based or historically observed values. This means that a systematic surplus is created by the company and, by law (which regulates mutual funds in some countries), it should be redistributed to the policyholder in terms of *bonuses* that are usually allocated but not distributed until the termination of the policy. Here we face a model risk!

• **Second order technical basis**: It is also called experience (or market) basis. It sets values of the parameters based on *realistic scenarios* collected based on the history of the policy. The company updates the reserves on a regular basis and adjusts for the parameters using the *bonus fund* created by applying the first order basis.

A typical example of adjustments to be made under the experience (market) basis is compensation for a possible non-equivalence of the first order payments i.e. $V_0(0^-) \neq 0$, i.e. the insurance company compensates for this by adding dividend payments D to the first order payments. D has usually the following form:

$$dD(t) := \sum_j \left(I_j(t)dD_j(t) + \sum_{k;k\neq j} \delta_{jk}(t)dN_{jk}(t) \right), \tag{20}$$

where,

$$dD_j(t) := \delta_j(t)dt + (D_j(t) - D_j(t^-)) = \delta_j(t)dt + \Delta D_j(t). \tag{21}$$

The coefficients $\delta_j, \Delta D_j$ and δ_{ij} are stochastic processes adapted to the "demographic-economic" history $\mathcal{F}$ with a more complex structure than the coefficients related to the payment processes A. The dividend process D is chosen (constrained) to attain the ultimate equivalence (fairness):

$$E\left[\int_{0^-}^{n} e^{-\int_0^s r(u)du} d(A + D)(s)\right] = 0. \tag{22}$$

In the Black and Scholes market model, the dividend payments are provided by an asset portfolio such as the following diffusion Y modulated by the jump process X:

$$\begin{aligned} &dY(t) = rY(t)dt + \sigma(t, X(t), Y(t))Y(t)dW(t) + d(C - D)(t), \\ &Y(0^-) = 0, \end{aligned} \tag{23}$$

where, C is the usual income (or contribution) process of the following form (similar to A and D):

$$dC(t) := \sum_j \left(I_j(t)dC_j(t) + \sum_{k;k\neq j} c_{jk}(t)dN_{jk}(t) \right), \tag{24}$$

$$dC_j(t) := c_j(t)dt + (C_j(t) - C_j(t^-)) = c_j(t)dt + \Delta C_j(t). \tag{25}$$

Assuming the coefficients $\delta_j(t)$, $\Delta D_j(t)$ and $\delta_{ij}(t)$ are functions of $(t, Y(t))$, the state-wise *prospective reserve* is

$$\begin{aligned} V_i(t,x) &:= E[V(t)|X(t)=i, Y(t)=x] \\ &= E\left[\int_t^n e^{-\int_t^s r(u)du} d(A+D)(s)|X(t)=i, Y(t)=x\right] \end{aligned} \tag{26}$$

satisfies a more complex "Thiele's" PDE (cf. [8, 16, 18]).

1.7 Graduation Techniques-Estimation of the Mortality Rates

We start with statistical inference of the mortality rate μ which is the only jump intensity in the simplest life insurance contract: Single life with one cause of death (one absorbing state) i.e. $E = \{0, 1\}$, where state 0 = alive, state 1= dead (absorbing state). The underlying Markov chain counts the number of deaths:

$$X(t) = \mathbb{1}_{\{T \leq t\}}, \quad t \in [0, n],$$

where, T denotes the life length of a person with survival probability

$$p_{00}(s,t) = \frac{\bar{F}(t)}{\bar{F}(s)} = e^{-\int_s^t \mu(u)du}, \qquad 0 \leq s \leq t \leq n.$$

In actuarial practice one often considers the *remaining life length* T_x of an insured of age x. The corresponding survival probability over a time period of length $t \geq 0$ is

$$P(T_x > t) := P(T > x+t|T > x) = e^{-\int_x^{x+t} \mu(u)du} = e^{-\int_0^t \mu(x+u)du}. \tag{27}$$

In a more general framework where 'stochastic mortality' modeling can be incorporated, consider (the possibly random) force (or rate) of mortality $\mu(x, t)$ at t for individual aged x at time 0. Then, the *survival index*

$$S(x,t) := \exp\left(-\int_0^t \mu(x+s, s)ds\right)$$

is the probability of survival of an individual aged x during the time interval $[0, t]$, given the mortality force $\mu(x, s)$ i.e.

$$P(T_x > t) = E[S(x,t)].$$

In Eq. (27), $\mu(x,t) = \mu(x+t)$.

The main goal of this section is to estimate the mortality force $\mu(x, s)$, given historical mortality data of a population of insured individuals.

1.7.1 An Age-Specific Model: Gompertz-Makeham Graduation Formula

This model captures the evolution of mortality in mutually exclusive age cohorts but disregards a possible common risk factor that links all cohorts together. Consider an insured population of ages $x_i,\ i = 1, 2, \ldots, n$. Let N_x denote the exposure i.e. the number of individuals of the same age x, and D_x denotes the number of individuals dead during the interval $(x, x + 1)$. Assuming that the remaining survival lengths of all individuals are *independent*, and the insured population is *homogeneous* in the sense that the survival probability of all individuals is the same. A stochastic model based on a "crude approximation" of the Binomial distribution by the Poisson distribution suggests that

$$D_{x_i} \sim \text{independent Poisson}(\mu_{x_i} N_{x_i}). \tag{28}$$

Then the mortality rate (or force) μ_{x_i} for a population of age $x_i,\ i = 1, 2, \ldots, n$ can be estimated by the so-called 'central or crude death rate'

$$\hat{\mu}_{x_i} = \frac{D_{x_i}}{N_{x_i}}, \quad i = 1, 2, \ldots, n. \tag{29}$$

Gompertz and later Makeham famous graduation formula suggests a mortality rate of the form

$$\mu_x := \alpha + \beta e^{\gamma x}, \tag{30}$$

where, the parameters α, β and γ which satisfy $\alpha + \beta > 0$, $\beta > 0$ and $\gamma \geq 0$ are estimated using the insured population data. When $\alpha = 0$ we get Gompertz mortality law. A fairly standard way to perform the parameter estimation is to use a weighted least squares method: minimize

$$Q = \sum_{i=0}^{n} w_{x_i} \left(\hat{\mu}_{x_i} - \alpha - \beta e^{\gamma x_i}\right)^2 \tag{31}$$

w.r.t. the parameters α, β and γ, where the weight is the inverse of the variance of $\hat{\mu}_{x_i}$:

$$w_{x_i} = \frac{N_{x_i}^2}{Var(D_{x_i})} = \frac{N_{x_i}^2}{N_{x_i}\hat{\mu}_{x_i}} = \frac{N_{x_i}}{\hat{\mu}_{x_i}}, \tag{32}$$

so that Q is approximately χ^2-distributed. In practice, one 'fixes' a value for γ 'based on experience' and finds the optimal values of α and β. In the Swedish life insurance business, there is a Central Mortality Committee that estimates these parameters to be used by insurance companies and pension funds. For example, in the so-called M90 investigation, the committee suggested that

$$\mu_x = \alpha + \beta e^{\gamma(x-f)},$$

where, the parameter f adjusts for mortality of females among the insured population. Values $f = 4$ or 5 years are used. For M90, $\alpha = 0.001$, $\beta = 0.000012$ and $\gamma = 0.044$.

1.7.2 Gompertz Graduation Formula with a View Towards GLM

Recall Gompertz' graduation formula:

$$\mu_x := \beta e^{\gamma x}, \tag{33}$$

or $\log \mu_x$, which is linear in age,

$$\log \mu_x = \log \beta + \log e^{\gamma x} := a + \gamma x.$$

This can be extended to a quadratic or a polynomial form

$$\log \mu_x = a + bx + cx^2, \quad \log \mu_x = a_0 + a_1 x + a_2 x^2 + \ldots + a_p x^p.$$

GLM means that we perform a regression of $\log \mu_x$ with respect to a basis

$$\{1, x\}, \quad \{1, x, x^2\}, \quad \{1, x, \ldots, x^p\},$$

or any other carefully chosen 'spline' basis $\{B_1(x), B_2(x), \ldots, B_p(x)\}$ such that

$$\mu_x = \sum_{j=1}^{p} B_j(x) a_j := P(a),$$

and estimate the coefficients $a_0, a_1, \ldots, a_p$ which maximize the *penalized log-likelihood function*:

$$L(a) - \frac{1}{2}\lambda P(a), \tag{34}$$

where, $L(a)$ is the log- likelihood of the model

$$D_{x_i} \sim \text{independent Poisson}(\mu_{x_i} N_{x_i}), \quad i = 1, \ldots, n, \tag{35}$$

and $\lambda > 0$ is a smoothing parameter.

A similar approach can be applied to obtain a smooth *year (or period) specific* mortality: maximize the *penalized log-likelihood function*

$$L(\theta) - \frac{1}{2}\lambda P(\theta), \tag{36}$$

where, $L(\theta)$ is the log- likelihood of the model

$$D_{t_i} \sim \text{independent Poisson}(\mu_{t_i} N_{t_i}), \quad t = t_{\min}, \ldots, t_{\max}, \tag{37}$$

and

$$P(\theta) = \sum_{j=1}^{p} B_j(t)\theta_j.$$

The smoothing parameter λ can be estimated using the Akaike Information Criterion (AIC), the Bayesian Information Criterion (BIC) or the Generalized Cross-Validation (GCV).

1.7.3 An Age-Period Model: Lee-Carter Graduation Formula

Lee and Carter [10] suggest a Gompertz type graduation formula for the full mortality rate $\mu(x, t)$:

$$\log \mu(x, t) := \alpha(x) + \beta(x)\kappa(t), \tag{38}$$

subject to the constraints

$$\sum_x \beta(x) = 1, \quad \sum_t \kappa(t) = 0, \tag{39}$$

fitting

$$\sum_{x,t} (\log \mu_{obs}(x, t) - \alpha(x) + \beta(x)\kappa(t))^2 .$$

This model captures the evolution of mortality in mutually exclusive age cohorts while at the same time includes a possible common risk factor (systemic risk) $k(t)$ that links all cohorts together over time. The parameters $a(x)$ and $b(x)$ are age-specific while $k(t)$ is time (period) dependent only and should capture the random period effect of the mortality rate. The risk factor $k(t)$ is usually modeled as a time series or a random walk with drift. Lee and Carter [10] suggest an ARIMA (discretized diffusion process) for κ of the form

$$k(t+1) = k(t) + a_1 + a_2\xi + \sigma z(t)$$

where, $z(t)$ is white noise and $\xi \in \{0, 1\}$ is a dummy variable that captures major outbreaks of disease leading to a huge mortality wave such as the 1918 worldwide flu outbreak or the 2008 earthquake in China etc. Statistical estimation of these parameters is usually performed w.r.t. each dimension: x and time (period) t. Here are some suggestions (see [2, 10], Currie, Richards and co-authors (2003–2012), [17] etc.).

- Given $\kappa(t) = \hat{\kappa}(t)$, fit a GLM with regressor $\hat{\kappa}$:

$$\log \mu(x, t) := \alpha(x) + \beta(x)\hat{\kappa}(t).$$

- Given $\alpha(x) = \hat{\alpha}(x), \quad \beta(x) = \hat{\beta}(x)$, fit a GLM with offset $\hat{\alpha}(x)$ and regressor $\hat{\beta}(x)$:

$$\log \mu(x, t) := \hat{\alpha}(x) + \hat{\beta}(x)\kappa(t).$$

- Perform a regression w.r.t. a 2-d spline basis $B_a(x) \otimes B_y(t)$ for age and time dimensions (x, t).

1.7.4 Building Blocks of the MLE for the Lee-Carter Model

Following [2], the MLE approach to the Lee-Carter model is based on the assumption that

$$\begin{aligned} D_{x,t} \sim \text{Poisson}(\mu(x,t)N_{x,t}), \quad &\text{where } \log \mu(x,t) := \alpha(x) + \beta(x)\kappa(t), \\ x = x_{\min}, \ldots, x_{\max}, \quad &t = t_{\min}, \ldots, t_{\max}. \end{aligned} \tag{40}$$

The parameters $\alpha(x)$, $\beta(x)$ and $\kappa(t)$ are estimated by maximizing the log-likelihood function

$$L(\alpha, \beta, \kappa) := \sum_{x,t} \big(D_{x,t}(\alpha(x) + \beta(x)\kappa(t)) - N_{x,t} \exp\left(\alpha(x) + \beta(x)\kappa(t)\right)\big) + C,$$

where, C contains all the terms that do not dependent on the parameters. The nonlinear term $\beta(x)\kappa(t)$ does not allow for a closed form of the maximizing parameters. One instead uses an iterative method such as the Newton-Raphson updating scheme (or any more efficient numerical optimization algorithm):

$$\theta^{(n+1)} = \theta^{(n)} - \frac{\partial L^{(n)}/\partial\theta}{\partial^2 L^{(n)}/\partial\theta^2},$$

which numerically solves $\partial L^{(n)}/\partial\theta = 0$.

$$\begin{cases} \hat{\alpha}_x^{(0)} = 0, \ \hat{\beta}_x^{(0)} = 1, \ \hat{\kappa}_t^{(0)} = 0, \\ \text{alternatively } \hat{\alpha}_x^{(0)} = \frac{1}{t_{\max}-t_{\min}+1} \sum_t \log(\hat{\mu}(x,t)), \ \hat{\beta}_x^{(0)} = \frac{1}{t_{\max}-t_{\min}+1}, \\ \hat{\kappa}_t^{(0)} = \sum_x \hat{\beta}_x^{(0)} \left(\log(\hat{\mu}(x,t)) - \hat{\alpha}_x^{(0)} \right), \\ \hat{D}_{x,t}^{(n)} = N_{x,t} \exp(\hat{\alpha}_x^{(n)} + \hat{\beta}^{(n)} \hat{\kappa}_t^{(n)}), \\ \hat{\alpha}_x^{(n+1)} = \hat{\alpha}_x^{(n)} - \frac{\sum_t (D_{x,t} - \hat{D}_{x,t}^{(n)})}{-\sum_t \hat{D}_{x,t}^{(n)}}, \quad \hat{\beta}_x^{(n+1)} = \hat{\beta}_x^{(n)}, \quad \hat{\kappa}_t^{(n+1)} = \hat{\kappa}_t^{(n)}, \\ \hat{\kappa}_t^{(n+2)} = \hat{\kappa}_t^{(n)} - \frac{\sum_t (D_{x,t} - \hat{D}_{x,t}^{(n+1)}) \hat{\beta}_x^{(n+1)}}{-\sum_t \hat{D}_{x,t}^{(n+1)} (\hat{\beta}_x^{(n+1)})^2}, \quad \hat{\alpha}_x^{(n+2)} = \hat{\alpha}_x^{(n+1)}, \ \hat{\beta}_x^{(n+2)} = \hat{\beta}_x^{(n+1)}, \\ \hat{\beta}_x^{(n+3)} = \hat{\beta}_x^{(n+2)} - \frac{\sum_t (D_{x,t} - \hat{D}_{x,t}^{(n+2)}) \hat{\kappa}^{n+2}(t)}{-\sum_t \hat{D}_{x,t}^{(n+2)} (\hat{\kappa}_t^{n+2})^2}, \quad \hat{\alpha}_x^{(n+3)} = \hat{\alpha}_x^{(n+2)}, \ \hat{\kappa}_t^{(n+3)} = \hat{\kappa}_t^{(n+2)}. \end{cases}$$

The parameters are standardized in each step of the iteration to satisfy the constraints

$$\sum_x \beta(x) = 1, \quad \sum_t \kappa(t) = 0, \tag{41}$$

by letting

$$\hat{\alpha}_x^{(n+1)} = \hat{\alpha}_x^{(n)} + A\hat{\beta}_x^{(n)}, \quad \hat{\kappa}_x^{(n+1)} = (\hat{\kappa}_t^{(n)} - A)B, \quad \hat{\beta}_x^{(n+1)} = \hat{\beta}_x^{(n)}/B, \tag{42}$$

where,

$$A = \frac{1}{t_{\max} - t_{\min}} \sum_t \hat{\kappa}_t^{(n)}, \quad B = \sum_x \hat{\beta}_x^{(n)}. \tag{43}$$

The estimated values of $\kappa(t)$, $t = t_{\min}, \ldots, t_{\max}$ are used to fit it to a dynamical model. We mentioned above that Lee and Carter fit $\kappa(t)$ to an ARIMA model of the form

$$k(t+1) = k(t) + a_1 + a_2\xi + \sigma z(t)$$

where, $z(t)$ is white noise and $\xi \in \{0, 1\}$ is a dummy variable that captures major mortality changes.

This algorithm is illustrated by the Figs. 4, 5 and 6, applied to mortality data among Swedish insured (cf. Swedish Research Board for Actuarial Science [17]).

Mortality jumps, due to e.g. new life standards or medical development etc., are also important to capture in a mortality model, despite the serious difficulties to perform reliable estimation. Cox et al. [7] suggest two types of mortality jump events to the Lee-Carter model:

$$\log \mu(x,t) := \alpha(x) + \beta(x)\kappa(t) - G(x,t) + H(x,t),$$

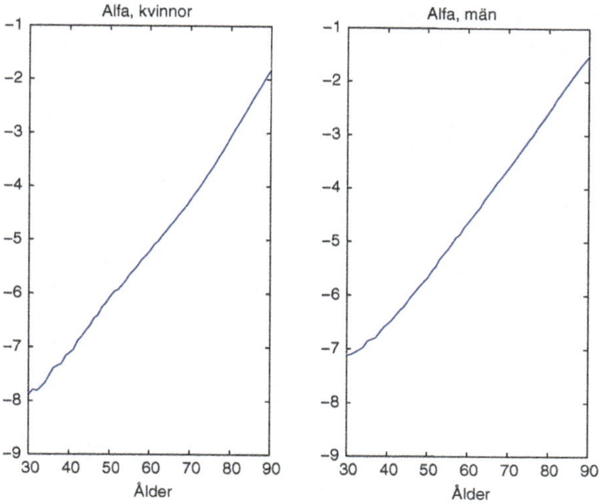

Fig. 4 The α_x parameter for ages 30–90 years (females and males). (From [17], reproduced with permission from Taylor and Francis Ltd.)

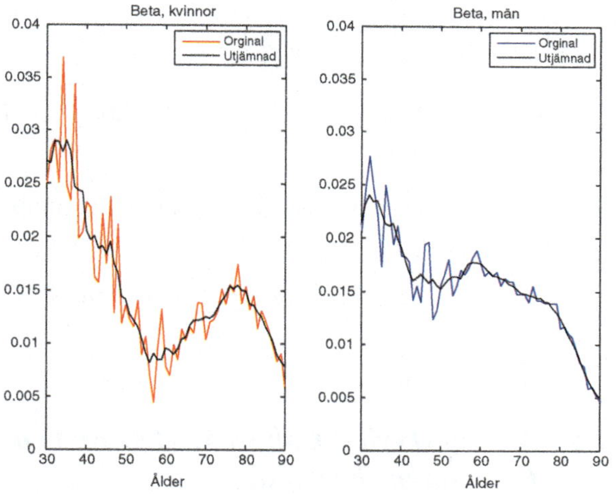

Fig. 5 Estimated and smoothed β_x parameter for ages 30–90 years (females and males). (From [17], reproduced with permission from Taylor and Francis Ltd.)

where

- $G(x, t)$ captures a *permanent longevity jump* and takes the form

$$G(x,t) := K(x,t) + D(x,t),$$

with

$$K(x,t) := \sum_{j=1}^{\infty} y_j A_j(x) \mathbb{1}_{\{t \geq \eta_j\}} = \text{Jump reduction component},$$

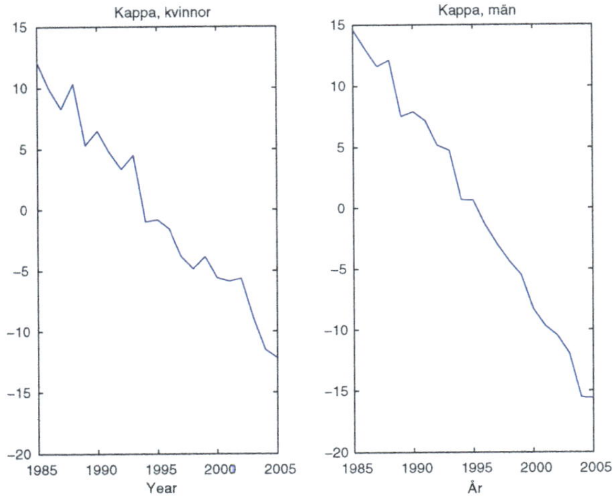

Fig. 6 Estimated and linearized $\kappa(t)$ parameter for data 1985–2005 (females and males). (From [17], reproduced with permission from Taylor and Francis Ltd.)

and

$$D(x,t) := \sum_{j=1}^{\infty} \zeta_i(t-\nu_i)F_i(x)e^{-\xi_i(t-\nu_i)}1\!\!1_{\{t\geq\nu_i\}} = \text{Trend reduction component.}$$

- $K(x,t)$ captures temporary adverse mortality jumps and takes the form

$$H(x,t) := \sum_{j=0}^{\infty} b_j B_j(x)e^{-\kappa_i(t-\tau_i)}1\!\!1_{\{t\geq\tau_j\}}.$$

1.8 An Age-Period-Cohort Model: Extending Lee-Carter Graduation Formula

The Lee-Carter model captures the age-period effect, but does not reflect the possible cohort effect (calender year-age $= t - x$). A simple model that would simultaneously capture the age-period-cohort effect is

$$\log \mu(x,t) := \alpha(x) + \kappa(t) + \gamma(t-x).$$

Renshaw and Haberman [9] suggested the following extension of the Lee-Carter model to capture the cohort effect (calender year-age $= t - x$):

$$\log \mu(x,t) := \beta_1(x) + \beta_2(x)\kappa(t) + \beta_3(x)\gamma(t-x). \tag{44}$$

A generalization of this mortality model for data divided into N components reads

$$\log \mu(x,t) := \sum_{j=1}^{N} \beta_j(x)\kappa_j(t)\gamma_j(t-x).$$

In a series of papers the Edinburgh teams including Currie, Richards and co-authors (2003–2012) and Cairns and co-authors (2006–2012) suggest other extensions and perform deep statistical analysis that seem tune the age-period-cohort effect when applied to mortality data from England and Wales, and USA.

1.9 An Infinite Dimensional Approach to Mortality Modeling

The mortality rate can be viewed as an (infinite dimensional) curve of (x, t). To capture the high level of uncertainty in projections of future mortality one is tempted to translate the "machinery" developed for "forward" interest rate yields such as "the HJM-model under the Musiela parametrization etc." to mortality rates. One is tempted to translated the calibration techniques of interest rate yield curves, to perform hopefully more accurate projections of future mortality (though with limited data points). Recent relevant references include [3–6, 19].

2 Disability Insurance

In the next sections we briefly describe a stochastic semi-Markov model for the development of disability inception and recovery rates and perform the corresponding statistical estimation. For more details see [1].

2.1 Disability Inception

Let $E_{x,t}$ denote the number of healthy individuals with age in $[x, x+1)$ at the beginning of time period t, and let $D_{x,t}$ denote the number of individuals among $E_{x,t}$ with disability inception in the interval $[t, t+1)$. In this section we model inception over time, $t = 0, 1, 2, \ldots$ and eventually estimate the underling parameters.

The Fig. 7 describes inception frequencies per 5-year age groups of females insured and a smoothed curve. This plot clearly shows that inception seems to be strongly time- and age-dependent. Below, we suggest a model of this behavior.

Assume $D_{x,t}$ is binomially distributed given $E_{x,t}$:

$$D_{x,t} \sim \text{Bin}(E_{x,t}, p_{x,t}) \tag{45}$$

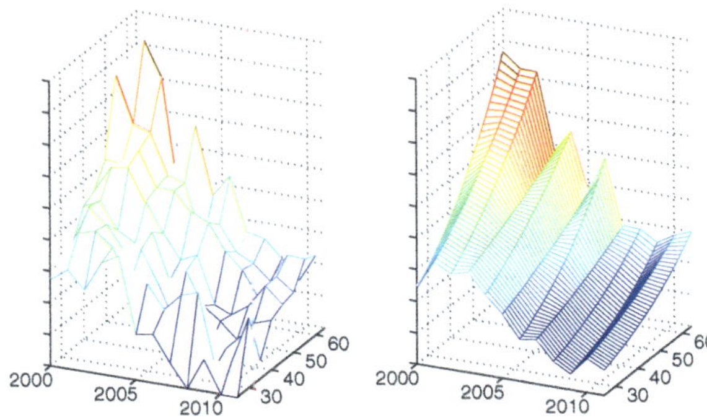

Fig. 7 *Left* Inception frequencies per 5-year age groups, females. *Right* Smoothed surface

where $p_{x,t}$ is the inception probability of an x-year-old. In order to reduce the dimensionality of the problem and achieve some level of smoothness, we use the logistic regression:

$$\operatorname{logit} p_{x,t} := \log\left(\frac{p_{x,t}}{1-p_{x,t}}\right) = \sum_{i=1}^{n} \nu_t^i \phi^i(x), \tag{46}$$

where $\phi^i(x)$ are age-dependent *basis functions*, and ν_t^i time-varying stochastic *risk factors* that we aim at estimating. Changing notation, $p_{\nu_t}(x) = p_{x,t}$, we invert the expression above, obtaining

$$p_{\nu_t}(x) = \frac{\exp\left(\sum_{i=1}^{n} \nu_t^i \phi^i(x)\right)}{1+\exp(\sum_{i=1}^{n} \nu_t^i \phi^i(x))}. \tag{47}$$

This guarantees that the probabilities $p_{\nu_t}(x) \in (0, 1)$.

Given historical values of $D_{x,t}$ and $E_{x,t}$, and a set of basis functions $\{\phi^i\}$, the log-likelihood function for yearly values of $\nu_t \in \mathbf{R}^n$ can be written

$$l(\nu_t) = \sum_{x\in X}\left[D_{x,t}\sum_{i=1}^{n} \nu_t^i \phi^i(x) - E_{x,t}\log\left(1+\exp\left\{\sum_{i=1}^{n} \nu_t^i \phi^i(x)\right\}\right)\right] + c_t. \tag{48}$$

If the basis functions are linearly independent it can be shown that $-l(\nu_t)$ is strictly convex. Thus it has a unique minimum. Minimizing $-l(\nu_t)$, using e.g. methods from numerical optimization, yields estimates of ν_t. The basis functions can be chosen by the user, according to some criteria. Desired properties of $p_{\nu_t}(\cdot)$, e.g. continuity or smoothness w.r.t. x, are achieved by choosing continuous or smooth $\phi^i(\cdot)$, by taking into account eventual population characteristics. Suitable choices of basis functions give the risk factors concrete interpretations. Alternatively, an optimal basis can be extracted from the data using functional principal component analysis. This approach yields better model fit, but harder to interpret results.

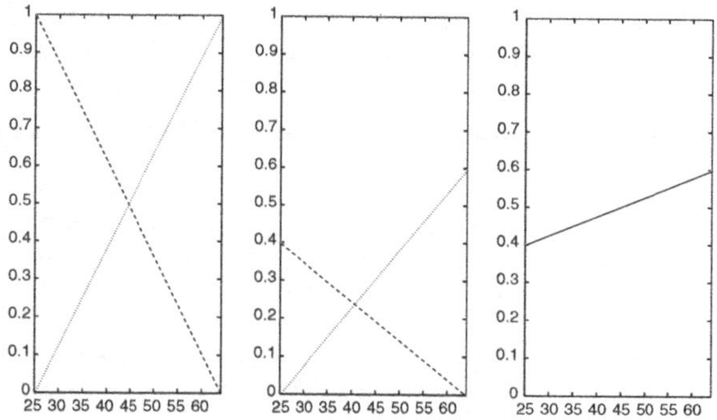

Fig. 8 *Left* Two basis functions. *Centre* Basis functions scaled with risk factor values 0.4 and 0.6. *Right* The resulting linear combination. *Note* $\phi^1(25) = \phi^2(64) = 1$, and $\phi^1(64) = \phi^2(25) = 0$

Consider the simple model

$$\operatorname{logit} p_{\nu_t}(x) = \nu_t^1 \phi^1(x) + \nu_t^2 \phi^2(x),$$

where the basis functions are linear on $x \in [25, 64]$:

$$\phi^1(x) = \frac{64 - x}{39} \quad , \quad \phi^2(x) = \frac{x - 25}{39}. \tag{49}$$

A linear combination of ϕ^1 and ϕ^2 is then also linear Fig. 8.
Under this model, the logistic inception probability of a 25-year old is given by

$$\operatorname{logit} p_{\nu_t}(25) = \nu_t^1 \phi^1(25) + \nu_t^2 \phi^2(25) = \nu_t^1.$$

Similarly, for a 64-year old we have $\operatorname{logit} p_{\nu_t}(64) = \nu_t^2$. An x-year old can be seen as a convex combination of a 25-year old and a 64-year old. Inception for the population is fully described by only ν_t^1 and ν_t^2.

2.2 *Recovery from Disability*

Recovery from disability is slightly more complicated. The probability of recovering from illness depends on the amount of time spent in the 'ill' state. This is known as the semi-Markov property. We extend the disability inception model above to the semi-Markov case, and apply it to recovery modeling.

Let $E_{x,d,t}$ denote the number of individuals with disability inception age in $[x, x+1)$ and disability duration d at some point in the time period $[t, t+1)$. Let $R_{x,d,t}$ denote the number of individuals among $E_{x,d,t}$ that recover during $[d, d+\Delta d)$ and $[t, t+1)$. Assume $R_{x,d,t}$ is binomially distributed given $E_{x,d,t}$:

$$R_{x,d,t} \sim \operatorname{Bin}(E_{x,d,t}, p_{x,d,t}), \tag{50}$$

where, $p_{x,d,t}$ is the probability that an individual, with disability inception age in $[x, x+1)$ and disability duration d at some point in $[t, t+1)$, recovers during $[d, d+\Delta d)$.

We propose the following logistic regression model:

$$\text{logit} p_{\nu_t}(x,d) = \sum_{i=1}^{n} \phi^i(x) \sum_{j=1}^{k} \nu_t^{i,j} \psi^j(d), \tag{51}$$

where ϕ^i and ψ^j, are age and duration dependent basis functions, respectively, and $\nu_t^{i,j}$ are stochastic risk factors. This is the inception model Eq. (46), extended with one dimension. The likelihood also has the same structure as before. It is strict convexity if each of the sets of functions $\{\phi^i\}$ and $\{\psi^i\}$ are linearly independent. Again, we estimate ν_t using numerical optimization Fig. 9.

Consider the simple model

$$\text{logit}\, p_{\nu_t}(x,d) = \phi^1(x) \sum_{j=1}^{3} \nu_t^{1,j} \psi^j(d) + \phi^2(x) \sum_{j=1}^{3} \nu_t^{2,j} \psi^j(d)$$

where ϕ and ψ are given by:

$$\phi^1(x) = \tfrac{64-x}{39},\ \phi^2(x) = \tfrac{x-25}{39},\ \psi^1(d) = 1,\ \psi^2(d) = d,\ \psi^3(d) = \sqrt{d}.$$

Hence, the recovery probabilities for a 25-year old are given by

$$\text{logit}\, p_{\nu_t}(25,\cdot) = \phi^1(25) \sum_{j=1}^{3} \nu_t^{1,j} \psi^j(\cdot) + \phi^2(25) \sum_{j=1}^{3} \nu_t^{2,j} \psi^j(\cdot) = \sum_{j=1}^{3} \nu_t^{1,j} \psi^j(\cdot),$$

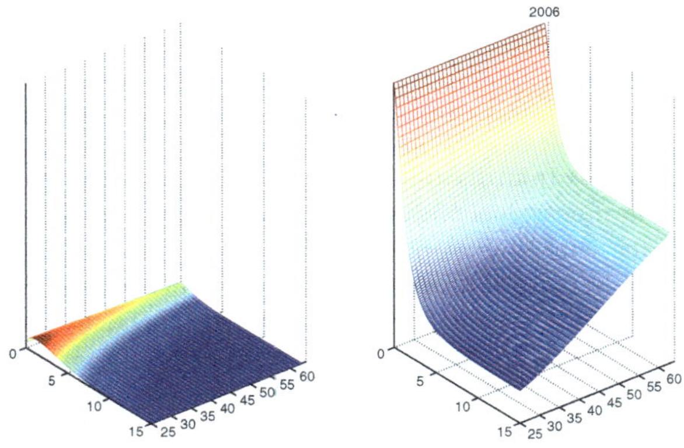

Fig. 9 *Left* Conditional recovery probabilities. *Right* Recovery surface, females, calendar year 2006

determined by $\nu_t^{1,1}$, $\nu_t^{1,2}$, $\nu_t^{1,3}$. Similarly, the recovery probabilities for a 64-year old determined by $\nu_t^{2,1}$, $\nu_t^{2,2}$, $\nu_t^{2,3}$. An x-year old can be seen as a convex combination of a 25-year old and a 64-year old. These considerations allow us to fully compute the probability that illness lasts longer than a given period. Let an x-year old's illness duration be the r.v. D. The probability that the illness lasts longer than d years is given by

$$\lambda(x, d) = P_{\nu_t}(D > d) = \prod_{n=0}^{d/\Delta d-1} (1 - p_{\nu_t}(x, n\Delta d)).$$

This is analogous to survival curves.

References

1. Aro, H., Djehiche, B., Löfdahl, B.: Stochastic modelling of disability insurance in a multi-period framework. Scand. Actuar. J. (2013). http://dx.doi.org/10.1080/03461238.2013.779594
2. Brouhns, N., Denuit, D., Vermunt, J.K.: A Poisson log-bilinear regression approach to the construction of projected lifetables. Insurance: Mathematics and Economics **31**(3), 373–393 (2002)
3. Biffis, E.: Affine processes for dynamic mortality and actuarial valuation. Insur. Math. Econom. **37**(3), 443–468 (2005)
4. Biffis, E., Denuit, M.: Lee-Carter goes risk-neutral: An application to the Italian annuity market, pp. 33–53. Giornale dell'Istituto Italiano degli Attuari, LXIX (2006)
5. Biffis, E., Denuit, M., Devolder, P.: Stochastic mortality under measure changes. Scand. Actua. J. **4**, 284–311 (2010)
6. Biffis, E., Millosovitch, P.: A bidimensional approach to mortality risk. Decision in Economics and Finance **29**(2), 71–94 (2006)
7. Cox, S.H., Lin, Y., Pedersen, H.: Mortality risk modeling: Applications to insurance securitization. Insurance: Mathematics and Economics **46**(1), 242–253 (2010)
8. Fahrenwaldt, M.A.: Sensitivity of life insurance reserves via Markov semigroups (preprint) (2013)
9. Haberman, S., Renshaw, A.: On age-period-cohort parametric mortality rate projections. Insurance Math. Econom. **45**(2), 255–270 (2009)
10. Lee, R.D., Carter, L.: Modelling and forecasting the time series of US mortality. Journal of the American Statistical Association **87**, 659–671 (1992)
11. Norberg, R.: Reserves in life and pension insurance. Scand. Actuar. J. **1991**(1), 324 (1991)
12. Norberg, R.: Identities for present values of life insurance. Scand. Actuar. J. pp. 100–106 (1993)
13. Norberg, R.: A theory of bonus prognoses in life insurance. Finance Stoch. **3**(4), 373–390 (1999)
14. Norberg, R.: On bonus and bonus prognoses in life insurance. Scand. Actuar. J. **2001**(2), 126–147 (2001)
15. Norberg, R.: Basic life insurance mathematics. University of Copenhagen, Lecture Notes (2002)
16. Persson, S.V.: Stochastic interest rate in life insurance: The principle of equivalence revisited. Scandinavian Actuarial Journal **1998**(2), 97–112 (1998)
17. Samuelsson, E.: Mortality among Swedish insured. Scan. Actua. J. **2008**(2–3), 184–199 (2008). www.tandfonline.com
18. Steffensen, M.: Surplus-liked life insurance. Scandinavian Actuarial Journal **1**, 1–22 (2006)
19. Tappe, S., Weber, S.: Stochastic mortality models: An infinite-dimensional approach (preprint) (2013)

Necessary and Sufficient Conditions of Optimalcontrol for Infinite Dimensional SDEs

Abdulrahman Al-Hussein

Abstract A general maximum principle (necessary and sufficient conditions) for an optimal control problem governed by a stochastic differential equation driven by an infinite dimensional martingale is established. The solution of this equation takes its values in a separable Hilbert space and the control domain need not be convex when studying optimality necessary conditions. The result is obtained by using the adjoint backward stochastic differential equation.

Keywords Martingale · Optimal control · Backward stochastic differential equation · Maximum principle · Conditions of optimality

Mathematical Subject Classification 2010 60H10 · 60G44

1 Introduction

This paper studies the following form of a controlled stochastic differential equation (SDE in short):

$$\begin{cases} dX(t) = F(X(t), u(t))dt + G(X(t))dM(t), \quad 0 \leq t \leq T, \\ X(0) = x_0, \end{cases} \tag{1}$$

where M is a continuous martingale taking its values in a separable Hilbert space K, while F, G are some mappings with properties to be given later and $u(\cdot)$ represents a control variable. We will be interested in minimizing the cost functional:

$$J(u(\cdot)) = \mathbb{E}\left[\int_0^T \ell(X^{u(\cdot)}(t), u(t))\, dt + h(X^{u(\cdot)}(T))\right]$$

over a set of admissible controls.

A. Al-Hussein (✉)
Department of Mathematics, College of Science, Qassim University, P.O. Box 6644, Buraydah 51452, Saudi Arabia
e-mail: alhusseinqu@hotmail.com; hsien@qu.edu.sa

M. Eddahbi et al. (eds.), *Statistical Methods and Applications in Insurance and Finance*, Springer Proceedings in Mathematics & Statistics 158,
DOI 10.1007/978-3-319-30417-5_6

We shall follow mainly the ideas of Bensoussan in [10, 11], Zhou in [36, 37], Øksendal et al. [26], and our earlier work [4]. The reader can see our main results in Theorems 2 and 3.

We recall that forward SDEs driven by martingales are studied in [6, 15, 16, 21, 34]. In fact in [6] we derived the maximum principle (necessary conditions) for optimality of stochastic systems governed by SPDEs. However, the results there show the maximum principle in its local form and also the control domain is assumed to be convex. In this paper we shall try to avoid such conditions as we shall shortly talk about it. Due to the fact that we are dealing here with a non-convex domain of controls, it is not obvious how one can allow the control variable $u(t)$ to enter in the mapping G in (1) and obtain a result like Lemma 3 below. This issue was raised also in [10]. Nevertheless, in some special cases (see [8]) we can allow G to depend on the control, still overcome this difficulty, and prove the maximum principle. The general case is still open as pointed out in [6, Remark 6.4].

The maximum principle in infinite dimensions started after the work of Pontryagin [30]. The reader can find a detailed description of these aspects in Li and Yong [22] and the references therein. An expanded discussion on the history of maximum principle can be found in [36, P. 153–156]. On the other hand, the use of (linear) backward stochastic differential equations (BSDEs) for deriving the maximum principle for forward controlled stochastic equations was done by Bismut in [12]. In this respect, one can see also the works of Bensoussan in [10, 11]. In 1990 Pardoux and Peng [27], initiated the theory of nonlinear BSDEs, and then Peng studied the stochastic maximum principle in [28, 29]. Since then several works appeared consequently on the maximum principle and its relationship with BSDEs. For example one can see [17–19, 33, 36] and the references of Zhou cited therein. Our earlier work in [2] has now opened the way to study BSDEs and backward SPDEs that are driven by martingales. One can see [23] for financial applications of BSDEs driven by martingales, and [7, 9, 14, 20] for other applications.

In this paper we shall consider first a suitable perturbation of an optimal control by means of the spike variation method in order to derive the maximum principle in its global form to derive optimality necessary conditions. Then we shall provide sufficient conditions for optimality of our control problem. The results will be achieved mainly by using the adjoint equation of (1), which is a BSDE driven by the martingale M. This can be seen from Eq. (30) in Sect. 5. It is quite important to realize that the adjoint equations in Sect. 5 of such SDEs are in general BSDEs driven by martingales. This happens also even if the martingale M, which is appearing in Eq. (1), is a Brownian motion with respect to a right continuous filtration being larger than its natural filtration. There is a discussion on this issue in Bensoussan's lecture note [10, Sect. 4.4], and in [1] and its erratum, [5]. In particular, studying control problems associated with SDEs like (1) with their martingale noises cannot be recovered from the works done for SDEs driven by Brownian motions in the literature. We refer the reader to the discussion at the beginning of Sect. 5 below for more details. To the best of our knowledge our results here towards deriving the maximum principle (necessary and sufficient optimality conditions) in its global form for a control problem governed by SDE (1) with a martingale noise are new. The general case when the control variable enters in the noise term G is still an open problem as stated above.

The paper is organized as follows. Section 2 is devoted to some preliminary notation. In Sect. 3 we present our main stochastic control problems. Then in Sect. 4 we establish many of our necessary estimates, which will be needed to derive the maximum principle for the control problem of (1). The maximum principle in the sense of Pontryagin for the above control problem is derived in Sect. 5. In Sect. 6 we establish sufficient conditions for optimality for this control problem, and present some examples as well.

2 Preliminary Notation

Let $(\Omega, \mathscr{F}, \mathbb{P})$ be a complete probability space, filtered by a continuous filtration $\{\mathscr{F}_t\}_{t\geq 0}$, in the sense that every square integrable K-valued martingale with respect to $\{\mathscr{F}_t,\ 0 \leq t \leq T\}$ has a continuous version.

Denoting by $\mathscr{P}$ the predictable σ—algebra of subsets of $\Omega \times [0, T]$ we say that a K - valued process is *predictable* if it is $\mathscr{P}/\mathscr{B}(K)$ measurable. Suppose that $\mathscr{M}^2_{[0,T]}(K)$ is the Hilbert space of cadlag square integrable martingales $\{M(t), 0 \leq t \leq T\}$, which take their values in K. Let $\mathscr{M}^{2,c}_{[0,T]}(K)$ be the subspace of $\mathscr{M}^2_{[0,T]}(K)$ consisting of all continuous square integrable martingales in K. Two elements M and N of $\mathscr{M}^2_{[0,T]}(K)$ are said to be *very strongly orthogonal* (or shortly VSO) if

$$\mathbb{E}\,[M(\tau) \otimes N(\tau)] = \mathbb{E}\,[M(0) \otimes N(0)],$$

for all $[0, T]$—valued stopping times τ.

Now for $M \in \mathscr{M}^{2,c}_{[0,T]}(K)$ we shall use the notation $< M >$ to mean the predictable quadratic variation of M and similarly $\ll M \gg$ means the predictable tensor quadratic variation of M, which takes its values in the space $L_1(K)$ of all nuclear operators on K. Precisely, $M \otimes M - \ll M \gg \in \mathscr{M}^{2,c}_{[0,T]}(L_1(K))$. We shall assume for a given fixed $M \in \mathscr{M}^{2,c}_{[0,T]}(K)$ that there exists a measurable mapping $\mathscr{Q}(\cdot) : [0, T] \times \Omega \to L_1(K)$ such that $\mathscr{Q}(t)$ is symmetric, positive definite, $\mathscr{Q}(t) \leq \mathscr{Q}$ for some positive definite nuclear operator $\mathscr{Q}$ on K, and satisfies the following equality:

$$\ll M \gg_t = \int_0^t \mathscr{Q}(s)\, ds.$$

We refer the reader to Example 1 for a precise computation of this process $\mathscr{Q}(\cdot)$.

For fixed (t, ω), we denote by $L_{\mathscr{Q}(t,\omega)}(K)$ to the set of all linear operators $\varphi : \mathscr{Q}^{1/2}(t, \omega)(K) \to K$ and satisfy $\varphi \mathscr{Q}^{1/2}(t, \omega) \in L_2(K)$, where $L_2(K)$ is the space of all Hilbert-Schmidt operators from K into itself. The inner product and norm in $L_2(K)$ will be denoted respectively by $\langle \cdot, \cdot \rangle_2$ and $|| \cdot ||_2$. Then the stochastic integral $\int_0^{\cdot} \Phi(s) dM(s)$ is defined for mappings Φ such that for each (t, ω), $\Phi(t, \omega) \in$

$L_{\mathcal{Q}(t,\omega)}(K)$, $\Phi\mathcal{Q}^{1/2}(t,\omega)(h)\ \forall\ h \in K$ is predictable, and

$$\mathbb{E}\left[\int_0^T ||(\Phi\mathcal{Q}^{1/2})(t)||_2^2\, dt\right] < \infty.$$

Such integrands form a Hilbert space with respect to the scalar product $(\Phi_1, \Phi_2) \mapsto \mathbb{E}\,[\int_0^T \langle \Phi_1\mathcal{Q}^{1/2}(t), \Phi_2\mathcal{Q}^{1/2}(t)\rangle\, dt\,]$. Simple processes taking values in $L(K;K)$ are examples of such integrands. By letting $\Lambda^2(K;\mathscr{P}, M)$ be the closure of the set of simple processes in this Hilbert space, it becomes a Hilbert subspace. We have also the following isometry property:

$$\mathbb{E}\left[|\int_0^T \Phi(t)dM(t)|^2\right] = \mathbb{E}\left[\int_0^T ||\Phi(t)\mathcal{Q}^{1/2}(t)||_2^2\, ds\right] \tag{2}$$

for mappings $\Phi \in \Lambda^2(K;\mathscr{P}, M)$. For more details and proofs we refer the reader to [25].

On the other hand, we emphasize that the process $\mathcal{Q}(\cdot)$ will be play an important role in deriving the adjoint equation of the SDE (1) as it can be seen from Eqs. (29), (30) in Sect. 5. This is due to the fact that the integrand Φ is not necessarily bounded. More precisely, it is needed in order for the mapping $\nabla_x H$, which appear in both equations, to be defined on the space $L_2(K)$, since the process $Z^{u(\cdot)}$ there need not be bounded. This always has to be considered when working with BSDEs or BSPDEs driven by infinite dimensional martingales.

Next let us introduce the following space:

$$L^2_{\mathscr{F}}(0,T;E) := \{\psi\colon [0,T]\times\Omega \to E,\ \text{predictable and } \mathbb{E}\left[\int_0^T |\psi(t)|^2 dt\right] < \infty\},$$

where E is a separable Hilbert space.

Since $\mathcal{Q}(t) \le \mathcal{Q}$ for all $t \in [0,T]$ a.s., it follows from [3, Proposition 2.2] that if $\Phi \in L^2_{\mathscr{F}}(0,T;L_{\mathcal{Q}}(K))$ (where as above $L_{\mathcal{Q}}(K)) = L_2(\mathcal{Q}^{1/2}(K);K))$, the space of all Hilbert-Schmidt operators from $\mathcal{Q}^{1/2}(K)$ into K), then $\Phi \in \Lambda^2(K;\mathscr{P}, M)$ and

$$\mathbb{E}\left[\int_0^T ||\Phi(t)\mathcal{Q}^{1/2}(t)||_2^2\, dt\right] \le \mathbb{E}\left[\int_0^T ||\Phi(t)||^2_{L_{\mathcal{Q}}(K)}\, dt\right]. \tag{3}$$

An example of such a mapping Φ is the mapping G in Eq. (1); see the domain of G in the introduction of the following section.

3 Formulation of the Control Problem

Let $\mathcal{O}$ be a separable Hilbert space and U be a nonempty subset of $\mathcal{O}$. We say that $u(\cdot)$: $[0, T] \times \Omega \to \mathcal{O}$ is *admissible* if $u(\cdot) \in L^2_{\mathcal{F}}(0, T; \mathcal{O})$ and $u(t) \in U \;\; a.e., \; a.s.$ The set of admissible controls will be denoted by $\mathcal{U}_{ad}$.

Let F: $K \times \mathcal{O} \to K$, G: $K \to L_{\mathcal{Q}}(K)$, ℓ: $K \times \mathcal{O} \to \mathbb{R}$ and h: $K \to \mathbb{R}$ be measurable mappings. Consider the following SDE:

$$\begin{cases} dX(t) = F(X(t), u(t))\, dt + G(X(t))\, dM(t), \quad t \in [0, T], \\ X(0) = x_0 \in K. \end{cases} \tag{4}$$

If assumption (E1), which is stated below, holds, then (4) attains a unique solution in $L^2_{\mathcal{F}}(0, T; K)$. The proof of this fact can be gleaned from [31] or [32]. In this case we shall denote the solution of (4) by $X^{u(\cdot)}$.

Our assumptions are the following.

(E1) F, G, ℓ, h are continuously Fréchet differentiable with respect to x, F and ℓ are continuously Fréchet differentiable with respect to u, the derivatives F_x, F_u, G_x, ℓ_x, ℓ_u are uniformly bounded, and

$$|h_x|_{L(K;K)} \leq k\,(1 + |x|_K)$$

for some constant $k > 0$.

In particular, $|F_x|_{L(K,K)} \leq C_1$, $||G_x||_{L(K,L_{\mathcal{Q}}(K))} \leq C_2$, $|F_v|_{L(\mathcal{O},K)} \leq C_3$, for some positive constants C_i, $i = 1, 2, 3$, and similarly for ℓ.

(E2) ℓ_x satisfies Lipschitz condition with respect to u uniformly in x.

Consider now the *cost functional*:

$$J(u(\cdot)) := \mathbb{E}\left[\int_0^T \ell(X^{u(\cdot)}(t), u(t))\, dt + h(X^{u(\cdot)}(T))\right], \tag{5}$$

for $u(\cdot) \in \mathcal{U}_{ad}$.

The control problem here is to minimize (5) over the set $\mathcal{U}_{ad}$. Any $u^*(\cdot) \in \mathcal{U}_{ad}$ satisfying

$$J(u^*(\cdot)) = \inf\{J(u(\cdot)) : \; u(\cdot) \in \mathcal{U}_{ad}\} \tag{6}$$

is called an *optimal control*, and its corresponding solution $X^* := X^{u^*(\cdot)}$ to (4) is called an *optimal solution* of the stochastic optimal control problem (4)–(6). In this case the pair $(X^*, u^*(\cdot))$ in this case is called an *optimal pair*.

Remark 1 We mention here that the mappings F, G and ℓ in (4) and (5) can be taken easily to depend on time t with a similar proof as established in the following sections, but rather, having more technical computations.

Since this control problem has no constraints we shall deal generally with progressively measurable controls. However, for the case when there are final state constraints, one can mimic our results in Sects. 4, 5 and 6, and use Ekeland's variational principle in a similar way to [24, 28] or [36].

In the following section we shall begin with some variational method in order to derive our main variational inequalities that are necessary to establish the main result of Sect. 5.

4 Estimates

Let $(X^*, u^*(\cdot))$ be the given optimal pair. Let $0 \leq t_0 < T$ be fixed and $0 < \varepsilon < T - t_0$. Let v be a random variable taking its values in U, $\mathscr{F}_{t_0}$-measurable and $\sup_{\omega \in \Omega} |v(\omega)| < \infty$. Consider the following spike variation of the control $u^*(\cdot)$:

$$u_\varepsilon(t) = \begin{cases} u^*(t) & \text{if } t \in [0, T] \backslash [t_0, t_0 + \varepsilon] \\ v & \text{if } t \in [t_0, t_0 + \varepsilon]. \end{cases} \tag{7}$$

Let $X^{u_\varepsilon(\cdot)}$ denote the solution of the SDE (4) corresponding to $u_\varepsilon(\cdot)$. We shall denote it briefly by X_ε. Observe that $X_\varepsilon(t) = X^*(t)$ for all $0 \leq t \leq t_0$.

The following lemmas will be very useful in proving the main results of Sect. 5.

Lemma 1 *Let (E1) hold. Assume that $\{p(t),\ t_0 \leq t \leq T\}$ is the solution of the following linear equation:*

$$\begin{cases} dp(t) = F_x(X^*(t), u^*(t))\, p(t)\, dt + G_x(X^*(t))\, p(t)\, dM(t), t_0 < t \leq T, \\ p(t_0) = F(X^*(t_0), v) - F(X^*(t_0), u^*(t_0)). \end{cases} \tag{8}$$

Then

$$\sup_{t \in [t_0, T]} \mathbb{E}[\, |p(t)|^2\,] < C$$

for some positive constant C.

Proof With the help of (E1) apply Itô's formula to compute $|p(t)|^2$, and take the expectation. The required result follows then by using Gronwall's inequality.

Lemma 2 *Assuming (E1) we have*

$$\mathbb{E}\left[\sup_{t_0 \leq t \leq T} |X_\varepsilon(t) - X^*(t)|^2\right] = o(\varepsilon).$$

Proof For $t_0 \leq t \leq t_0 + \varepsilon$ one observes that

$$X_\varepsilon(t) - X^*(t) = \int_{t_0}^t [F(X_\varepsilon(s), v) - F(X^*(s), v)]\, ds$$
$$+ \int_{t_0}^t [F(X^*(s), v) - F(X^*(s), u^*(s))]\, ds + \int_{t_0}^t [G(X_\varepsilon(s)) - G(X^*(s))] dM(s), \tag{9}$$

or, in particular,

$$|X_\varepsilon(t) - X^*(t)|^2 \le 3\,(t-t_0)\int_{t_0}^t |F(X_\varepsilon(s),v) - F(X^*(s),v)|^2\,ds$$
$$+3\,(t-t_0)\int_{t_0}^t |F(X^*(s),v) - F(X^*(s),u^*(s))|^2\,ds$$
$$+3\,|\int_{t_0}^t [G(X_\varepsilon(s)) - G(X^*(s))]dM(s)|^2. \tag{10}$$

But Taylor expansion implies the three identities:

$$F(X_\varepsilon(s),v) - F(X^*(s),v)$$
$$= \int_0^1 F_x(X^*(s),u^*(s)+\lambda(X_\varepsilon(s)-X^*(s)))\,(X_\varepsilon(s)-X^*(s))\,d\lambda, \tag{11}$$

$$F(X^*(s),v) - F(X^*(s),u^*(s))$$
$$= \int_0^1 F_v(X^*(s),u^*(s)+\lambda(v-u^*(s)))\,(v-u^*(s))\,d\lambda, \tag{12}$$

and

$$G(X_\varepsilon(s)) - G(X^*(s)) = \int_0^1 G_x(X^*(s) \;+\lambda(X_\varepsilon(s)-X^*(s)))\,(X_\varepsilon(s)-X^*(s))\,d\lambda$$
$$=: \Phi(s)\ (\in L_{\mathcal{Q}}(K)). \tag{13}$$

Then, by using (13), the isometry property (2), (3) and (E1) we deduce that for all $t \in [t_0, t_0+\varepsilon]$,

$$\mathbb{E}\left[|\int_{t_0}^t \big(G(X_\varepsilon(s)) - G(X^*(s))\big)\,dM(s)|^2\right] = \mathbb{E}\left[|\int_{t_0}^t \Phi(s)dM(s)|^2\right]$$
$$= \mathbb{E}\left[\int_{t_0}^t ||\Phi(s)\mathcal{Q}^{1/2}(s)||_2^2\,ds\right]$$
$$\le \mathbb{E}\left[\int_{t_0}^t ||\Phi(s)||_{L_{\mathcal{Q}}(K)}^2\,ds\right]$$
$$= \mathbb{E}\left[\int_{t_0}^t ||\int_0^1 G_x(X^*(s)+\lambda(X_\varepsilon(s)-X^*(s)))\,(X_\varepsilon(s)-X^*(s))\,d\lambda||_{L_{\mathcal{Q}}(K)}^2\,ds\right]$$
$$\le \mathbb{E}\left[\int_{t_0}^t\int_0^1 ||G_x(X^*(s)+\lambda(X_\varepsilon(s)-X^*(s)))\,(X_\varepsilon(s)-X^*(s))||_{L_{\mathcal{Q}}(K)}^2\,d\lambda\,ds\right]$$
$$\le C_2\,\mathbb{E}\left[\int_{t_0}^t |X_\varepsilon(s)-X^*(s)|^2\,ds\right]. \tag{14}$$

Therefore, from (10), (11), (12), (E1) and (14), it follows evidently that

$$\begin{aligned}\mathbb{E}[\,|X_\varepsilon(t) - X^*(t)|^2\,] \;&\leq 3\;(C_1\,(t-t_0) + C_2)\int_{t_0}^{t} \mathbb{E}[\,|\,X_\varepsilon(s) - X^*(s)\,|^2\,]\,ds \\ &\quad + 3\,(t-t_0)\,C_3\int_{t_0}^{t} \mathbb{E}[\,|\,v - u^*(s)\,|^2\,]\,ds,\end{aligned}$$

for all $t \in [t_0, t_0 + \varepsilon]$.

Hence by using Gronwall's inequality we obtain

$$\mathbb{E}[\,|\,X_\varepsilon(t) - X^*(t)\,|^2\,] \leq 3\,C_3\,(t-t_0)\,e^{3\,(C_1\,(t-t_0)+C_2)(t-t_0)} \times \int_{t_0}^{t_0+\varepsilon} \mathbb{E}[\,|v - u^*(s)|^2\,]\,ds, \tag{15}$$

for all $t \in [t_0, t_0 + \varepsilon]$. Consequently,

$$\mathbb{E}\left[\int_{t_0}^{t_0+\varepsilon} |\,X_\varepsilon(t) - X^*(t)\,|^2\,dt\right] \leq 3\,C_3\,\varepsilon^2\,e^{3\,(C_1\,\varepsilon + C_2)\varepsilon} \times \int_{t_0}^{t_0+\varepsilon} \mathbb{E}[\,|v - u^*(s)|^2\,]\,ds. \tag{16}$$

It follows then from (10), (15), standard martingale inequalities, (14) and (16) that

$$\begin{aligned}&\mathbb{E}\left[\sup_{t_0 \leq t \leq t_0+\varepsilon} |X_\varepsilon(t) - X^*(t)|^2\right] \\ &\leq 3\,C_3\,[\,3\,(C_1\,\varepsilon + 4C_2)\,\varepsilon\,e^{3\,(C_1\,\varepsilon+C_2)\varepsilon} \;+\; 1\,]\,\varepsilon \int_{t_0}^{t_0+\varepsilon} \mathbb{E}[\,|v - u^*(s)|^2\,]\,ds.\end{aligned} \tag{17}$$

Next, for $t_0 + \varepsilon \leq t \leq T$, we have

$$\begin{aligned}X_\varepsilon(t) - X^*(t) &= X_\varepsilon(t_0 + \varepsilon) - X^*(t_0 + \varepsilon) \\ &\quad + \int_{t_0+\varepsilon}^{t} [F(X_\varepsilon(s), u^*(s)) - F(X^*(s), u^*(s))]\,ds \\ &\quad + \int_{t_0+\varepsilon}^{t} [G(X_\varepsilon(s)) - G(X^*(s))]dM(s).\end{aligned} \tag{18}$$

Thus by working as before and applying (15) we derive

$$\mathbb{E}\left[\int_{t_0+\varepsilon}^{T} |\,X_\varepsilon(t) - X^*(t)\,|^2\,dt\right] \leq 9\,C_3\,\varepsilon^2 e^{C_4(\varepsilon)} \int_{t_0}^{t_0+\varepsilon} \mathbb{E}[\,|v - u^*(s)|^2\,]\,ds$$

and

$$\mathbb{E}\left[\sup_{t_0+\varepsilon\le t\le T}|X_\varepsilon(t)-X^*(t)|^2\right] \le 27\,C_3\,\varepsilon\,e^{C_4(\varepsilon)}\,[\,1+((T-t_0-\varepsilon)\,C_1+4\,C_2)\,\varepsilon\,]$$
$$\times\int_{t_0}^{t_0+\varepsilon}\mathbb{E}[\,|v-u^*(s)|^2\,]\,ds, \qquad (19)$$

where $C_4(\varepsilon)=[3\,\varepsilon^2+3\,(T-t_0-\varepsilon)^2]\,C_1+(T-t_0+2\,\varepsilon)\,C_2$.

Now (17) and (19) imply that

$$\mathbb{E}\left[\sup_{t_0\le t\le T}|X_\varepsilon(t)-X^*(t)|^2\right]\le (C_5(\varepsilon)+C_6(\varepsilon))\int_{t_0}^{t_0+\varepsilon}\mathbb{E}[\,|v-u^*(s)|^2\,]\,ds,$$

with the constants

$$C_5(\varepsilon)=3\,C_3\,[\,3\,(C_1\,\varepsilon+4C_2)\,\varepsilon\,e^{3\,(C_1\,\varepsilon+C_2)\varepsilon}+1\,]\,\varepsilon$$

and

$$C_6(\varepsilon)=27\,C_3\,\varepsilon\,e^{C_4(\varepsilon)}\,[\,1+((T-t_0-\varepsilon)\,C_1+4\,C_2)\,\varepsilon\,].$$

This completes the proof.

Remark 2 We note that for $a.e.\ s$,

$$\frac{1}{\varepsilon}\int_s^{s+\varepsilon}\mathbb{E}[\,|\phi(X^*(t),u^*(t))-\phi(X^*(s),u^*(s))|^2\,]\,dt\to 0,\quad\text{as }\ \varepsilon\to 0,\qquad (20)$$

for $\phi=F,\ell$. Indeed, if for example, $\phi=F$, then we may argue as in (12) to see that

$$\frac{1}{\varepsilon}\int_s^{s+\varepsilon}\mathbb{E}[\,|F(X^*(t),u^*(t))-F(X^*(s),u^*(s))|^2\,]\,dt$$
$$=\frac{1}{\varepsilon}\int_s^{s+\varepsilon}\mathbb{E}\left[\,|\int_0^1 F_v(X^*(t),u^*(s)+\lambda(u^*(t)-u^*(s)))\,(u^*(t)-u^*(s))\,d\lambda|^2\,dt\right]$$
$$\le\frac{1}{\varepsilon}\int_s^{s+\varepsilon}\mathbb{E}[\,|u^*(t)-u^*(s)|^2\,]\,dt.\qquad (21)$$

But since $\int_0^T\mathbb{E}[\,|u^*(t)-u^*(s)|^2\,]\,dt<\infty$ (for fixed s), then, as it is well-known from measure theory (e.g. [13]), there exists a subset O of $[0,T]$ such that $\mathrm{Leb}([0,T]\setminus O)=0$ and the function $O\ni t\mapsto\mathbb{E}[\,|u^*(t)-u^*(s)|^2\,]$ is continuous. Thus, if $s\in O$, this function is continuous in a neighborhood of s, and so we have

$$\frac{1}{\varepsilon}\int_s^{s+\varepsilon}\mathbb{E}[\,|u^*(t)-u^*(s)|^2\,]\,dt\to 0,\quad\text{as }\ \varepsilon\to 0,$$

which by (21) implies (20) for $\phi=F$.

We will choose t_0 such that (20) holds for $\phi=F,\ell$. This assumption will be considered until the end of Sect. 5.

Lemma 3 *Assume (E1). Let*

$$\xi_\varepsilon(t) = \frac{1}{\varepsilon}\,(X_\varepsilon(t) - X^*(t)) - p(t), \quad t \in [t_0, T].$$

Then

$$\lim_{\varepsilon \to 0} \mathbb{E}[\,|\xi_\varepsilon(T)|^2\,] = 0.$$

Proof First note that, for $t_0 \le t \le t_0 + \varepsilon$,

$$\begin{aligned} d\xi_\varepsilon(t) &= \frac{1}{\varepsilon}\,[\,F(X_\varepsilon(t), v) - F(X^*(t), u^*(t)) - \varepsilon\, F_x(X^*(t), u^*(t))\, p(t)\,]dt \\ &\quad + \frac{1}{\varepsilon}\,[\,G(X_\varepsilon(t)) - G(X^*(t)) - \varepsilon\, G_x(X^*(t))\, p(t)\,]dM(t), \\ \xi_\varepsilon(t_0) &= -\left(F(X^*(t_0), v) - F(X^*(t_0), u^*(t_0))\right). \end{aligned}$$

Thus

$$\begin{aligned} \xi_\varepsilon(t_0 + \varepsilon) &= \frac{1}{\varepsilon}\int_{t_0}^{t_0+\varepsilon} [\,F(X_\varepsilon(s), v) - F(X^*(s), v)\,]\,ds \\ &\quad + \frac{1}{\varepsilon}\int_{t_0}^{t_0+\varepsilon} [\,F(X^*(s), v) - F(X^*(t_0), v)\,]\,ds \\ &\quad + \frac{1}{\varepsilon}\int_{t_0}^{t_0+\varepsilon} [\,F(X^*(t_0), u^*(t_0)) - F(X^*(s), u^*(s))\,]\,ds \\ &\quad + \frac{1}{\varepsilon}\int_{t_0}^{t_0+\varepsilon} [\,G(X_\varepsilon(s)) - G(X^*(s))\,]dM(s) \\ &\quad - \int_{t_0}^{t_0+\varepsilon} F_x(X^*(s), u^*(s)) p(s)\, ds - \int_{t_0}^{t_0+\varepsilon} G_x(X^*(s)) p(s) dM(s). \end{aligned}$$

By using (2), (3) and (E1) we deduce

$$\begin{aligned} \mathbb{E}[\,|\,\xi_\varepsilon(t_0+\varepsilon)\,|^2\,] &\le 6\, C_1\, \mathbb{E}[\sup_{t_0 \le t \le t_0+\varepsilon} |X_\varepsilon(t) - X^*(t)|^2\,] \\ &\quad + 6 \sup_{t_0 \le t \le t_0+\varepsilon} \mathbb{E}[\,|F(X^*(t), v) - F(X^*(t_0), v)|^2\,] \\ &\quad + \frac{6}{\varepsilon}\int_{t_0}^{t_0+\varepsilon} \mathbb{E}[\,|F(X^*(s), u^*(s)) - F(X^*(t_0), u^*(t_0))|^2\,]\,ds \\ &\quad + \frac{6\,C_2}{\varepsilon}\,\mathbb{E}\left[\sup_{t_0 \le t \le t_0+\varepsilon} |X_\varepsilon(t) - X^*(t)|^2\right] + 6\,(C_1 + C_2)\,\mathbb{E}\left[\int_{t_0}^{t_0+\varepsilon} |p(s)|^2\, ds\right]. \end{aligned} \tag{22}$$

But from (17)

$$\frac{1}{\varepsilon}\,\mathbb{E}\left[\sup_{t_0\le t\le t_0+\varepsilon} |X_\varepsilon(t) - X^*(t)|^2\right]$$
$$\le 3\,C_3\,[\,3\,(C_1\,\varepsilon + 4C_2)\,\varepsilon\,e^{3\,(C_1\,\varepsilon+C_2)\varepsilon} + 1\,]\int_{t_0}^{t_0+\varepsilon} \mathbb{E}[\,|v - u^*(s)|^2\,]\,ds \to 0 \tag{23}$$

as $\varepsilon \to 0$. Also as in (11), by applying (E1) and (15), one gets

$$\begin{aligned}
&\mathbb{E}[\,|F(X^*(t), v) - F(X^*(t_0), v)|^2]\\
&\quad = \mathbb{E}\left[\,|\int_0^1 F_x(X^*(t_0) + \lambda(X^*(t) - X^*(t_0)), v)(X^*(t) - X^*(t_0))\,d\lambda|^2\right.\\
&\quad \le C_1\,\mathbb{E}[\,|X^*(t) - X^*(t_0)|^2]\\
&\quad \le 3\,C_1\,C_3\,\varepsilon\,e^{3\,(C_1\,\varepsilon+C_2)\varepsilon}\int_{t_0}^{t_0+\varepsilon} \mathbb{E}[\,|v - u^*(s)|^2\,]\,ds \to 0 \quad \text{as } \varepsilon \to 0.
\end{aligned} \tag{24}$$

Thus, by applying Lemma 2, (24), (23), (20) and Lemma 1 in (22), we deduce

$$\mathbb{E}[\,|\,\xi_\varepsilon(t_0 + \varepsilon)\,|^2\,] \to 0 \quad \text{as } \varepsilon \to 0. \tag{25}$$

Let us now assume that $t_0 + \varepsilon \le t \le T$. In this case we have

$$\begin{aligned}
d\xi_\varepsilon(t) &= \frac{1}{\varepsilon}\,[\,F(X_\varepsilon(t), u^*(t)) - F(X^*(t), u^*(t)) - \varepsilon\,F_x(X^*(t), u^*(t))\,p(t)\,]dt\\
&\quad + \frac{1}{\varepsilon}\,[\,G(X_\varepsilon(t)) - G(X^*(t)) - \varepsilon\,G_x(X^*(t))\,p(t)\,]dM(t),
\end{aligned}$$

or, in particular, by setting

$$\tilde{\Phi}_\varepsilon(s) = \int_0^1 [\,G_x(X^*(s) + \lambda(X_\varepsilon(s) - X^*(s))) - G_x(X^*(s))\,]\,p(s)\,d\lambda,$$

we get

$$\begin{aligned}
\xi_\varepsilon(t) &= \xi_\varepsilon(t_0 + \varepsilon) + \int_{t_0+\varepsilon}^{t}\int_0^1 F_x(X^*(s) + \lambda(X_\varepsilon(s) - X^*(s)), u^*(s))\,\xi_\varepsilon(s)\,d\lambda\,ds\\
&\quad + \int_{t_0+\varepsilon}^{t}\int_0^1 G_x(X^*(s) + \lambda(X_\varepsilon(s) - X^*(s)))\,\xi_\varepsilon(s)\,d\lambda\,dM(s)\\
&\quad + \int_{t_0+\varepsilon}^{t}\int_0^1 [\,F_x(X^*(s) + \lambda(X_\varepsilon(s) - X^*(s)), u^*(s)) - F_x(X^*(s), u^*(s))\,]\,p(s)\,d\lambda\,ds\\
&\quad + \int_{t_0+\varepsilon}^{t} \tilde{\Phi}_\varepsilon(s)\,dM(s),
\end{aligned}$$

for all $t \in [t_0 + \varepsilon, T]$. Hence by making use of the isometry property (2) it holds $\forall\, t \in [t_0 + \varepsilon, T]$,

$$\mathbb{E}[\,|\,\xi_\varepsilon(t)\,|^2\,] \leq 5\,\mathbb{E}[\,|\,\xi_\varepsilon(t_0+\varepsilon)\,|^2\,] + 5\,(C_1 + C_2) \int_{t_0+\varepsilon}^{t} \mathbb{E}[\,|\xi_\varepsilon(s)\,|^2\,]\,ds$$
$$+5\,\mathbb{E}\left[\int_{t_0}^{T} \left|\int_0^1 \big(F_x(X^*(s)+\lambda(X_\varepsilon(s)-X^*(s)),u^*(s)) - F_x(X^*(s),u^*(s))\big)\,p(s)\,d\lambda\,ds\,\right|\right]^2$$
$$+5\,\mathbb{E}\left[\int_{t_0}^{T} ||\tilde{\Phi}_\varepsilon(s)\,\mathcal{Q}^{1/2}(s)||_2^2\,ds\right]. \tag{26}$$

But as done for the second equality and first inequality in (14) we can derive easily that

$$\mathbb{E}\left[\int_{t_0}^{T} ||\tilde{\Phi}_\varepsilon(s)\,\mathcal{Q}^{1/2}(s)||_2^2\,ds\right] = \mathbb{E}\left[\int_{t_0}^{t} ||\tilde{\Phi}_\varepsilon(s)\mathcal{Q}^{1/2}(s)||_2^2\,ds\right]$$
$$\leq \mathbb{E}\left[\int_{t_0}^{t} ||\tilde{\Phi}_\varepsilon(s)||^2_{L_{\mathcal{Q}}(K)}\,ds\right]$$
$$= \mathbb{E}\left[\int_{t_0}^{t} \left|\left|\int_0^1 [G_x(X^*(s)+\lambda(X_\varepsilon(s)-X^*(s))) - G_x(X^*(s))]p(s)\,d\lambda\right|\right|^2_{L_{\mathcal{Q}}(K)} ds\right]$$
$$\leq \mathbb{E}\left[\int_{t_0}^{t}\int_0^1 ||G_x(X^*(s)+\lambda(X_\varepsilon(s)-X^*(s))) - G_x(X^*(s))\,]\,p(s)||^2_{L_{\mathcal{Q}}(K)}\,d\lambda\,ds\right]. \tag{27}$$

Therefore, from Lemma 2, the continuity and boundedness of G_x in (E1), Lemma 1 and the dominated convergence theorem we deduce that the last term in the right hand side of (27) goes to 0 as $\varepsilon \to 0$.

Similarly, the third term in the right hand side of (26) converges also to 0 as $\varepsilon \to 0$.

Finally, by applying Gronwall's inequality to (26), and using (25)–(27), we deduce that

$$\sup_{t_0+\varepsilon \leq t \leq T} \mathbb{E}[\,|\,\xi_\varepsilon(t)\,|^2\,] \to 0 \;\text{ as }\; \varepsilon \to 0,$$

which proves the lemma.

Lemma 4 *Assume (E1) and (E2). Let ζ be the solution of the equation:*

$$\begin{cases} d\zeta(t) = \ell_x(X^*(t),u^*(t))p(t)dt, & t_0 < t \leq T, \\ \zeta(t_0) = \ell(X^*(t_0),v) - \ell(X^*(t_0),u^*(t_0)). \end{cases}$$

Then

$$\lim_{\varepsilon \to 0} \mathbb{E}\left[\left|\frac{1}{\varepsilon}\int_{t_0}^{T} \big(\ell(X_\varepsilon(t),u_\varepsilon(t)) - \ell(X^*(t),u^*(t))\big)\,dt - \zeta(t)\right|^2\right] = 0.$$

Proof Let

$$\eta_\varepsilon(t) = \frac{1}{\varepsilon}\int_{t_0}^{t} \big(\ell(X_\varepsilon(t), u_\varepsilon(t)) - \ell(X^*(t), u^*(t))\big)\,dt - \zeta(T),$$

for $t \in [t_0, T]$. Then $\eta_\varepsilon(t_0) = -\,(\ell(X^*(t_0), v) - \ell(X^*(t_0), u^*(t_0)))$. So one can proceed easily as done in the proof of Lemma 3 to show that $\mathbb{E}[\,|\,\eta_\varepsilon(T)\,|^2\,] \to 0$, though this case is rather simpler.

Let us now for a C^1 mapping $\Psi : K \to \mathbb{R}$ denote by $\nabla\Psi$ to the gradient of Ψ, which is defined, by using the directional derivative $D\Psi(x)(k)$ of Ψ at a point $x \in K$ in the direction of $k \in K$, as $\langle \nabla\Psi(x), k\rangle = D\Psi(x)(k)\ (= \Psi_x(k))$. We shall sometimes write $\nabla_x\Psi$ for $\nabla\Psi(x)$.

Corollary 1 *Under the assumptions of Lemma 4*

$$\frac{d}{d\varepsilon} J(u_\varepsilon(\cdot))|_{\varepsilon=0} = \mathbb{E}[\,\langle \nabla h(X^*(T)),\ p(T)\rangle + \zeta(T)\,]. \tag{28}$$

Proof Note that from the definition of the cost functional in (5) we see that

$$\begin{aligned}
\frac{1}{\varepsilon}\,\big[J(u_\varepsilon(\cdot)) - J(u^*(\cdot))\big] &= \frac{1}{\varepsilon}\,\mathbb{E}\,\Big[h(X_\varepsilon(T)) - h(X^*(T)) \\
&\quad + \int_{t_0}^{T} \big(\ell(X_\varepsilon(s), u_\varepsilon(s)) - \ell(X^*(s), u^*(s))\big)\,ds\Big] \\
= \mathbb{E}\Big[\int_0^1 h_x(X^*(T) &+ \lambda(X_\varepsilon(T) - X^*(T)))\,\frac{(X_\varepsilon(T) - X^*(T))}{\varepsilon}\,d\lambda \\
&\quad + \frac{1}{\varepsilon}\int_{t_0}^{T} \big(\ell(X_\varepsilon(s), u_\varepsilon(s)) - \ell(X^*(s), u^*(s))\big)\,ds\Big].
\end{aligned}$$

Now let $\varepsilon \to 0$ and use the properties of h in (E1), Lemmas 3 and 4 to get (28).

5 Maximum Principle

The maximum principle is a good tool for studying the optimality of controlled SDEs like (4) since in fact the dynamic programming approach for similar optimal control problems require usually a Markov property to be satisfied by the solution of (4), cf. for instance [36, Chap. 4]. But this property does not hold in general especially when the driving noise is a martingale.

Let us recall the SDE (4) and the mappings in (5), and define the *Hamiltonian* $H : [0, T] \times \Omega \times K \times \mathcal{O} \times K \times L_2(K) \to \mathbb{R}$ for $(t, \omega, x, u, y, z) \in [0, T] \times \Omega \times K \times \mathcal{O} \times K \times L_2(K)$ by

$$H(t,\omega,x,u,y,z) := \ell(x,u) + \langle F(x,u), y\rangle + \langle G(x)\mathcal{Q}^{1/2}(t,\omega), z\rangle_2 . \quad (29)$$

The adjoint equation of (4) is the following BSDE:

$$\begin{cases} -dY^{u(\cdot)}(t) = \nabla_x H(t, X^{u(\cdot)}(t), u(t), Y^{u(\cdot)}(t), Z^{u(\cdot)}(t)\mathcal{Q}^{1/2}(t))\,dt \\ \qquad\qquad - Z^{u(\cdot)}(t)\,dM(t) - dN^{u(\cdot)}(t), \quad t_0 \leq t < T, \\ Y^{u(\cdot)}(T) = \nabla h(X^{u(\cdot)}(T)). \end{cases} \quad (30)$$

The following theorem gives the solution to BSDE (30) in the sense that there exists a triple $(Y^{u(\cdot)}, Z^{u(\cdot)}, N^{u(\cdot)})$ in $L^2_{\mathcal{F}}(0,T;K) \times \Lambda^2(K;\mathcal{P},M) \times \mathcal{M}^{2,c}_{[0,T]}(K)$ such that the following equality holds $a.s.$ for all $t \in [0,T]$, $N(0) = 0$ and N is VSO to M:

$$Y^{u(\cdot)}(t) = \xi + \int_t^T \nabla_x H(s, X^{u(\cdot)}(s), u(s), Y^{u(\cdot)}(s), Z^{u(\cdot)}(s)\mathcal{Q}^{1/2}(s))\,ds \\ - \int_t^T Z^{u(\cdot)}(s)dM(s) - \int_t^T dN^{u(\cdot)}(s).$$

Theorem 1 *Assume that (E1)–(E2) hold. Then there exists a unique solution $(Y^{u(\cdot)}, Z^{u(\cdot)}, N^{u(\cdot)})$ of the BSDE (30).*

For the proof of this theorem one can see [2].

We shall denote briefly the solution of (30), which corresponds to the optimal control $u^*(\cdot)$ by (Y^*, Z^*, N^*).

In the following lemma we shall try to compute $\mathbb{E}[\,\langle Y^*(T), p(T)\rangle\,]$.

Lemma 5

$$\mathbb{E}[\,\langle Y^*(T), p(T)\rangle\,] = -\mathbb{E}\left[\int_{t_0}^T \ell_x(X^*(s), u^*(s))p(s)\,ds\right] \\ + \mathbb{E}\left[\langle Y^*(t_0), F(X^*(t_0), v) - F(X^*(t_0), u^*(t_0))\rangle\right]. \quad (31)$$

Proof Use Itô's formula together to compute $d\,\langle Y^*(t), p(t)\rangle$ for $t \in [t_0, T]$, and use the facts that

$$\int_{t_0}^T \langle\, p(s), \nabla_x H(s, X^*(s), u^*(s), Y^*(s), Z^*(s)\mathcal{Q}^{1/2}(s))\rangle\,ds \\ = \int_{t_0}^T \left[\ell_x(X^*(s), u^*(s))p(s) + \langle\, F_x(X^*(s), u^*(s))p(s), Y^*(s)\,\rangle\right] ds \\ + \int_{t_0}^T \langle\, G_x(X^*(s))p(s)\mathcal{Q}^{1/2}(s), Z^*(s)\mathcal{Q}^{1/2}(s)\,\rangle_2\,ds,$$

which is easily seen from (29).

Now we state our main result of this section.

Theorem 2 *Suppose (E1)–(E2). If $(X^*, u^*(\cdot))$ is an optimal pair for the problem (4)–(6), then there exists a unique solution (Y^*, Z^*, N^*) to the corresponding BSDE (30) such that the following inequality holds:*

$$H(t, X^*(t), v, Y^*(t), Z^*(t)\mathcal{Q}^{1/2}(t)) \geq H(t, X^*(t), u^*(t), Y^*(t), Z^*(t)\mathcal{Q}^{1/2}(t)) \quad a.e.\ t \in [0, T],\ a.s.\ \forall\, v \in U. \tag{32}$$

Proof We note that since $u^*(\cdot)$ is optimal, $\frac{d}{d\varepsilon} J(u_\varepsilon(\cdot))|_{\varepsilon=0} \geq 0$, which implies by using Corollary 1 that

$$\mathbb{E}[\, \langle Y^*(T),\ p(T)\rangle + \zeta(T)\,] \geq 0. \tag{33}$$

On other hand by applying (33) and Lemma 5 one sees that

$$0 \leq -\,\mathbb{E}[\int_{t_0}^{T} \ell_x(X^*(s), u^*(s))p(s)\,ds\,] + \mathbb{E}[\,\langle Y^*(t_0), F(X^*(t_0), v) - F(X^*(t_0), u^*(t_0))\rangle + \zeta(T)\,]. \tag{34}$$

But

$$\zeta(T) = \zeta(t_0) + \int_{t_0}^{T} \ell_x(X^*(s), u^*(s))p(s)\,ds$$

and

$$\begin{aligned} &H(t_0, X^*(t_0), v, Y^*(t_0), Z^*(t_0)\mathcal{Q}^{1/2}(t_0)) \\ &\quad - H(t_0, X^*(t_0), u^*(t_0), Y^*(t_0), Z^*(t_0)\mathcal{Q}^{1/2}(t_0)) \\ &\quad = \zeta(t_0) + \langle Y^*(t_0), F(X^*(t_0), v) - F(X^*(t_0), u^*(t_0))\rangle. \end{aligned}$$

Hence (34) becomes

$$\begin{aligned} 0 \leq \mathbb{E}\,[\, &H(t_0, X^*(t_0), v, Y^*(t_0), Z^*(t_0)\mathcal{Q}^{1/2}(t_0)) \\ &- H(t_0, X^*(t_0), u^*(t_0), Y^*(t_0), Z^*(t_0)\mathcal{Q}^{1/2}(t_0))\,]. \end{aligned} \tag{35}$$

Now varying t_0 as in (20) shows that (35) holds for $a.e.\ t.$, and so by arguing for instance as in [10, p. 19] we obtain easily (32).

Remark 3 Let us assume for example that the space K in Theorem 1 is the real space $\mathbb{R}$ and M is the martingale given by the formula

$$M(t) = \int_0^t \alpha(s)dB(s), \qquad t \in [0, T],$$

for some $\alpha \in L^2_{\mathscr{F}}(0, T; \mathbb{R})$ and a one dimensional Brownian motion B. If $\alpha(s) > 0$ for each s, then $\mathscr{F}_t(M) = \mathscr{F}_t(B)$ for each t, where

$$\mathscr{F}_t(R) = \sigma\{R(s), 0 \leq s \leq t\}$$

for $R = M, B$. Consequently, by applying the unique representation property for martingales with respect to $\{\mathscr{F}_t(M),\ t \geq 0\}$ or larger filtration in [2, Theorem 2.2] or [5] and the Brownian martingale representation theorem as e.g. in [14, Theorem 3.4, P. 200], we deduce that the martingale $N^{u(\cdot)}$ in (30) vanishes almost surely if the filtration furnished for the SDE (4) is $\{\mathscr{F}_t(M),\ 0 \leq t \leq T\}$. This result follows from the construction of the solution of the BSDE (30). More details on this matter can be found in [2, Sect. 3]. As a result, in this particular case BSDE (30) fits well with those BSDEs studied by Pardoux & Peng in [27], but with the variable αZ replacing Z there.

Thus in particular we conclude that many of the applications of BSDEs, which were studied in the literature, to both stochastic optimal control and finance (e.g. [37] and the references therein) can be applied directly or after slight modification to work here for BSDEs driven by martingales. For example we refer the reader to [23] for financial application. Another interesting case can be found in [9].

On the other hand, in this respect we shall present an example (see Example 2) in Sect. 6, by modifying an interesting example due to Bensoussan [10].

6 Sufficient Conditions for Optimality

In the previous two sections we derived Pontyagin's maximum principle which gives necessary conditions for optimality for the control problem (4)–(6). In the following theorem we shall try to show when the necessary conditions for optimality becomes sufficient as well. Let us assume from here on that U is convex. This concerned result is a variation of Theorem 4.2 in [3].

Theorem 3 *Assume (E1) and, for a given $u^*(\cdot) \in \mathscr{U}_{ad}$, let X^* and (Y^*, Z^*, N^*) be the corresponding solutions of Eqs. (4) and (30) respectively. Suppose that the following conditions hold:*

(i) U is a convex domain in $\mathscr{O}$, h is convex.
(ii) $(x, v) \mapsto H(t, x, v, Y^*(t), Z^*(t)\mathscr{Q}^{1/2}(t))$ is convex for all $t \in [0, T]$ a.s.
(iii) $H(t, X^*(t), u^*(t), Y^*(t), Z^*(t)\mathscr{Q}^{1/2}(t))$

$$= \min_{v \in U} H(t, X^*(t), v, Y^*(t), Z^*(t)\mathscr{Q}^{1/2}(t))$$

for a.e. $t \in [0, T]$ a.s.

Then $(X^*, u^*(\cdot))$ is an optimal pair for the control problem (4)–(6).

Proof Let $u(\cdot) \in \mathscr{U}_{ad}$. Consider the following definitions:

$$I_1 := \mathbb{E}\left[\int_0^T \big(\ell(X^*(t), u^*(t)) - \ell(X^{u(\cdot)}(t), u(t))\big)\, dt\right]$$

and

$$I_2 := \mathbb{E}\,[\, h(X^*(T)) - h(X^{u(\cdot)}(T))\,].$$

Then readily

$$J(u^*(\cdot)) - J(u(\cdot)) = I_1 + I_2. \tag{36}$$

Let us define

$$\begin{aligned} I_3 := \mathbb{E}\,[\int_0^T & \big(H(t, X^*(t), u^*(t), Y^*(t), Z^*(t)\, \mathscr{Q}^{1/2}(t)) \\ & - H(t, X^{u(\cdot)}(t), u(t), Y^*(t), Z^*(t)\, \mathscr{Q}^{1/2}(t))\big)dt\ ], \end{aligned}$$

$$I_4 := \mathbb{E}\left[\int_0^T \langle F(X^*(t), u^*(t)) - F(X^{u(\cdot)}(t), u(t)), Y^*(t)\rangle\, dt\right],$$

$$I_5 := \mathbb{E}\left[\int_0^T \langle \big(G(X^{u^*(\cdot)}(t)) - G(X^{u(\cdot)}(t))\big)\, \mathscr{Q}^{1/2}(t), Z^*(t)\mathscr{Q}^{1/2}(t)\rangle_2\, dt\right],$$

and

$$I_6 := \mathbb{E}\left[\int_0^T \langle \nabla_x H(t, X^*(t), u^*(t), Y^*(t), Z^{u^*(\cdot)}(t)\mathscr{Q}^{1/2}(t)), X^*(t) - X^{u(\cdot)}(t)\rangle\, dt\right].$$

From the definition of H in (29) we get

$$I_1 = I_3 - I_4 - I_5. \tag{37}$$

On the other hand, from the convexity of h in condition (ii) it follows

$$h(X^*(T)) - h(X^{u(\cdot)}(T)) \leq \langle \nabla h(X^*(T)), X^*(T) - X^{u(\cdot)}(T)\rangle \ \ a.s.,$$

which implies that

$$I_2 \leq \mathbb{E}\,[\,\langle\, Y^*(T), X^*(T) - X^u(T)\,\rangle\,]. \tag{38}$$

Next by applying Itô's formula to compute $d\,\langle Y^*(t), X^*(t) - X^{u(\cdot)}(t)\rangle$ and using Eqs. (4) and (30) we find with the help of (38) that

$$I_2 \leq I_4 + I_5 - I_6\,. \tag{39}$$

Consequently, by considering (36), (37) and (39) it follows that

$$J(u^*(\cdot)) - J(u(\cdot)) \leq I_3 - I_6. \tag{40}$$

On the other hand, from the convexity property of the mapping $(x, v) \mapsto H(t, x, u, Y^*(t), Z^*(t)\mathscr{Q}^{1/2}(t))$ in assumption (iii) the following inequality holds a.s.:

$$\begin{aligned}
&\int_0^T \Big(H(t, X^*(t), u^*(t), Y^*(t), Z^*(t)\mathscr{Q}^{1/2}(t)) \\
&\qquad\qquad - H(t, X^{u(\cdot)}(t), u(t), Y^*(t), Z^*(t)\mathscr{Q}^{1/2}(t))\Big)\, dt \\
&\leq \int_0^T \langle\, \nabla_x H(t, X^*(t), u^*(t), Y^*(t), Z^*(t)\mathscr{Q}^{1/2}(t)),\; X^*(t) - X^{u(\cdot)}(t)\,\rangle\, dt \\
&+ \int_0^T \langle\, \nabla_u H(t, X^*(t), u^*(t), Y^*(t), Z^*(t)\mathscr{Q}^{1/2}(t)),\; u^*(t) - u(t)\,\rangle_{\mathscr{O}}\, dt.
\end{aligned}$$

As a result

$$I_3 \leq I_6 + I_7, \tag{41}$$

where

$$I_7 = \mathbb{E}\left[\int_0^T \langle\, \nabla_u H(t, X^*(t), u^*(t), Y^*(t), Z^*(t)\mathscr{Q}^{1/2}(t)), u^*(t) - u(t)\,\rangle_{\mathscr{O}}\, dt\right].$$

Since $v \mapsto H(t, X^*(t), v, Y^*(t), Z^*(t)\mathscr{Q}^{1/2}(t))$ is minimum at $v = u^*(t)$ by the minimum condition (iii), we have

$$\langle\, \nabla_u H(t, X^*(t), u^*(t), Y^*(t), Z^*(t)\mathscr{Q}^{1/2}(t)), u^*(t) - u(t)\,\rangle_{\mathscr{O}} \leq 0.$$

Therefore $I_7 \leq 0$, which by (41) implies that $I_3 - I_6 \leq 0$. So (40) becomes

$$J(u^*(\cdot)) - J(u(\cdot)) \leq 0.$$

Now since $u(\cdot) \in \mathscr{U}_{ad}$ is arbitrary, this inequality proves that $(X^*, u^*(\cdot))$ is an optimal pair for the control problem (4)–(6) as required.

Example 1 Let m be a continuous square integrable one dimensional martingale with respect to $\{\mathscr{F}_t\}_t$ such that $< m >_t = \int_0^t \alpha(s)ds \;\forall\; 0 \leq t \leq T$ for some continuous $\alpha : [0, T] \to (0, \infty)$. Consider $M(t) = \beta\, m(t) (= \int_0^t \beta\, dm(s))$, with $\beta \neq 0$ being a fixed element of K. Then $M \in \mathscr{M}^{2,c}(K)$ and $\ll M \gg_t$ equals $\widetilde{\beta \otimes \beta} \int_0^t \alpha(s)ds$, where $\widetilde{\beta \otimes \beta}$ is the identification of $\beta \otimes \beta$ in $L_1(K)$, that is $(\widetilde{\beta \otimes \beta})(k) = \langle \beta, k \rangle \beta,\ k \in K$. Also $< M >_t = |\beta|^2 \int_0^t \alpha(s)\, ds$. Now letting $\mathscr{Q}(t) = \widetilde{\beta \otimes \beta}\, \alpha(t)$ yields that $\ll M \gg_t = \int_0^t \mathscr{Q}(s)\, ds$. This process $\mathscr{Q}(\cdot)$ is bounded since $\mathscr{Q}(t) \leq \mathscr{Q}\ \ \forall\, t$, where $\mathscr{Q} = \widetilde{\beta \otimes \beta} \max_{0 \leq t \leq T} \alpha(t)$. It is also easy to see that $\mathscr{Q}^{1/2}(t)(k) = \frac{\langle \beta, k \rangle \beta}{|\beta|} \alpha^{1/2}(t)$. In particular $\beta \in \mathscr{Q}^{1/2}(t)(K)$.

Let $K = L^2(\mathbb{R}^n)$. Let M be the above martingale. Suppose that $\mathscr{O} = K$. Assume that $\tilde{G} \in L_{\mathscr{Q}}(K)$ or even a bounded linear operator from K into itself, and $\tilde{F}$ is a bounded linear operator from $\mathscr{O}$ into K. Let us consider the SDE:

$$\begin{cases} dX(t) = \tilde{F}\, u(t)\, dt + \langle X(t), \beta \rangle\, \tilde{G}\, dM(t), \quad t \in [0, T], \\ X(0) = x_0 \in K. \end{cases}$$

For a given fixed element c of K we assume that the cost functional is given by the formula:

$$J(u(\cdot)) = \mathbb{E}\left[\int_0^T |u(t)|^2\, dt \right] + \mathbb{E}\left[\, \langle c, X(T) \rangle \,\right],$$

and the value function is

$$J^* = \inf\{ J(u(\cdot)) :\ u(\cdot) \in \mathscr{U}_{ad} \}.$$

This control problem can be related to the control problem (4)–(6) as follows. We define

$$F(x, u) = \tilde{F}\, u, \ \ G(x) = \langle x, \beta \rangle\, \tilde{G}, \ \ \ell(x, u) = |u|^2, \ \text{ and } \ h(x) = \langle c, x \rangle,$$

where $(x, u) \in K \times \mathscr{O}$.

The Hamiltonian then becomes the mapping

$$H : [0, T] \times \Omega \times K \times \mathscr{O} \times K \times L_2(K) \to \mathbb{R},$$

$$H(t, x, u, y, z) = |u|^2 + \langle \tilde{F}\, u, y \rangle + \langle x, \beta \rangle\, \langle \tilde{G}\, \mathscr{Q}^{1/2}(t), z \rangle_2,$$

$(t, x, u, y, z) \in K \times \mathscr{O} \times K \times L_2(K)$.

It is obvious that $H(\cdot, \cdot, y, z)$ is convex with respect to (x, u) for each y and z and $\nabla_x H(t, x, u, y, z) = \langle \tilde{G}\, \mathscr{Q}^{1/2}(t), z \rangle\, \beta$.

Next we consider the adjoint BSDE:

$$\begin{cases} -dY(t) = [\, \langle \tilde{G}\, \mathscr{Q}^{1/2}(t), Z(t) \rangle_2\, \beta \,]\, dt - Z(t)\, dM(t) - dN(t), \\ Y(T) = c. \end{cases}$$

This BSDE attains an explicit solution $Y(t) = c$, since c is non-random. But this implies that $Z(t) = 0$ and $N(t) = 0$ for each $t \in [0, T]$.

On the other hand, we note that the function $\mathcal{O} \ni u \mapsto H(t, x, u, y, z) \in \mathbb{R}$ attains its minimum at $u = -\frac{1}{2} \tilde{F}^* y$, for fixed (x, y, z). So we choose our candidate for an optimal control as

$$u^*(t, \omega) = -\frac{1}{2} \tilde{F}^* Y(t, \omega) = -\frac{1}{2} \tilde{F}^* c \quad (\in U := \mathcal{O}).$$

With this choice all the requirements in Theorem 3 are verified. Consequently $u^*(\cdot)$ is an optimal control of this control problem with an optimal solution $\hat{X}$ given by the solution of the following closed loop equation:

$$\begin{cases} d\hat{X}(t) = -\frac{1}{2} \tilde{F} \tilde{F}^* Y(t)\, dt + \langle \hat{X}(t), \beta \rangle\, \tilde{G}\, dM(t), \\ \hat{X}(0) = x_0 \in K. \end{cases}$$

The value function takes the following value:

$$J^* = \frac{1}{4} |\tilde{F}^* c|^2\, T + \mathbb{E}\, [\, \langle c, \hat{X}(T) \rangle\,].$$

Remark 4 It would be possible if we take $h(x) = |x|^2$, $x \in K$, in the preceding example and proceeds as above. However if a result of existence and uniqueness os solutions to what we may call "forward-backward stochastic differential equations with martingale noise" holds, it should certainly be very useful to deal with both this particular case and similar problems.

Example 2 Let $\mathcal{O} = K$. We are interested in the following linear quadratic example, which is gleaned from Bensoussan [10, p. 33]. Namely, we consider the SDE:

$$\begin{cases} dX(t) = (A(t)X(t) + C(t)u(t) + f(t))\, dt + (B(t)X(t) + D(t))\, dM(t), \\ X(0) = x_0, \end{cases} \tag{42}$$

where $B(t)x = \langle \gamma(t), x \rangle\, \tilde{G}(t)$ and $A, \gamma, C : [0, T] \times K \to K$, $f : [0, T] \to K$, $\tilde{G}, D : [0, T] \to L_{\mathcal{Q}}(K)$ are measurable and bounded mappings.

Let $P, Q : [0, T] \times K \to K$, $P_1 : K \to K$ be measurable and bounded mappings. Assume that P, P_1 are symmetric non-negative definite, and Q is a symmetric positive definite and $Q^{-1}(t)$ is bounded. For SDE (42) we shall assume that the cost functional is

$$\begin{aligned} J(u(\cdot)) &= \mathbb{E} \left[\int_0^T \left(\frac{1}{2} \langle P(t)X^{u(\cdot)}(t), X^{u(\cdot)}(t) \rangle + \frac{1}{2} \langle Q(t)u(t), u(t) \rangle \right) dt \right. \\ &\quad \left. + \frac{1}{2} \langle P_1 X^{u(\cdot)}(T), X^{u(\cdot)}(T) \rangle \right], \end{aligned} \tag{43}$$

for $u(\cdot) \in \mathcal{U}_{ad}$.

The control problem now is to minimize (43) over the set $\mathscr{U}_{ad}$ and get an optimal control $u^*(\cdot) \in \mathscr{U}_{ad}$, that is

$$J(u^*(\cdot)) = \inf\{J(u(\cdot)) : u(\cdot) \in \mathscr{U}_{ad}\}. \tag{44}$$

By recalling Remark 1 we can consider this control problem (42)–(44) as a control problem of the type (4)–(6). To this end, we let

$$F(t, x, u) = A(t)x + C(t)u + f(t),$$

$$G(t, x) = \langle \gamma(t), x\rangle \, \tilde{G}(t) + D(t),$$

$$\ell(t, x, u) = \frac{1}{2} \langle P(t)x, x\rangle + \frac{1}{2} \langle Q(t)u, u\rangle,$$

$$h(x) = \frac{1}{2} \langle P_1 x, x\rangle.$$

Then the Hamiltonian $H : [0, T] \times \Omega \times K \times K \times K \times L_2(K) \to \mathbb{R}$ is given by

$$\begin{aligned} H(t, x, u, y, z) &= \ell(t, x, u) + \langle F(t, x, u), y\rangle + \langle G(t, x)\mathscr{Q}^{1/2}(t), z\rangle_2 \\ &= \frac{1}{2} \langle P(t)x, x\rangle + \frac{1}{2} \langle Q(t)u, u\rangle + \langle A(t)x + C(t)u + f(t), y\rangle \\ &\quad + \langle \, (\langle \gamma(t), x\rangle \, \tilde{G}(t) + D(t))\mathscr{Q}^{1/2}(t), z\rangle_2 \, . \end{aligned}$$

We can compute $\nabla_x H$ directly to find that

$$\nabla_x H(t, x, u, y, z) = P(t)u + A^*(t)x + \langle \tilde{G}(t)\mathscr{Q}^{1/2}(t), z\rangle_2 \, \gamma(t).$$

Hence the adjoint equation of (42) takes the following shape:

$$\begin{cases} -dY^{u(\cdot)}(t) = \Big(A^*(t)Y^{u(\cdot)}(t) + P(t)X^{u(\cdot)}(t) \\ \qquad\qquad + \big\langle \tilde{G}(t)\mathscr{Q}^{1/2}(t), Z^{u(\cdot)}(t)\mathscr{Q}^{1/2}(t)\big\rangle_2 \, \gamma(t) \Big) \, dt \\ \qquad\qquad - Z^{u(\cdot)}(t)dM(t) - dN^{u(\cdot)}(t), \\ Y^{u(\cdot)}(T) = P_1 X^{u(\cdot)}(T). \end{cases}$$

Now the maximum principle theorems (Theorems 2, 3) in this case hold readily if we consider Remark 1 again, and yield eventually

$$C^*(t)Y^*(t) + \frac{1}{2} \, Q(t)u^*(t) = 0.$$

Acknowledgments The author would like to thank the Mathematics Institute, Warwick University, where part of this work was done, for hospitality during the summer of 2011. Many thanks and much gratitude to King Abdulaziz City for Science and Technology (KACST), Riyadh, Saudi Arabia, for their support.

References

1. Al-Hussein, A.: Backward stochastic partial differential equations in infinite dimensions. Random Oper. Stoch. Equ. **14**(1), 1–22 (2006)
2. Al-Hussein, A.: Backward stochastic partial differential equations driven by infinite dimensional martingales and applications. Stochastics **81**(6), 601–626 (2009)
3. Al-Hussein, A.: Maximum principle for controlled stochastic evolution equations. Int. J. Math. Anal. **4**(30), 1447–1464 (2010)
4. Al-Hussein, A.: Sufficient conditions of optimality for backward stochastic evolution equations. Commun. Stoch. Anal. **4**(3), 433–442 (2010)
5. Al-Hussein, A.: Representation of infinite dimensional martingales. Random Oper. Stoch. Equ. **18**, 127–130 (2010)
6. Al-Hussein, A.: Necessary conditions for optimal control of stochastic evolution equations in Hilbert spaces. Appl. Math. Optim. **63**(3), 385–400 (2011)
7. Al-Hussein, A.: BSDEs driven by infinite dimensional martingales and their applications to stochastic optimal control. Random Oper. Stoch. Equ. **19**, 45–61 (2011)
8. Al-Hussein, A.: Maximum principle for optimal control of infinite dimensional stochastic differential equations. Dyn. Syst. Appl. **21**, 205–218 (2012)
9. Bally, V., Pardoux, É., Stoica, L.: Backward stochastic differential equations associated to a symmetric Markov process. Potential Anal. **22**(1), 17–60 (2005)
10. Bensoussan, A.: Lectures on stochastic control. Nonlinear Filtering and Stochastic Control, Cortona, 1981. Lecture Notes in Math., vol. 972, pp. 1–62. Springer, Berlin (1982)
11. Bensoussan, A.: Stochastic Control of Partially Observable Systems. Cambridge University Press, Cambridge (1992)
12. Bismut, J.-M.: Théorie probabiliste du contrôle des diffusions. Mem. Amer. Math. Soc. **4**, 167 (1976)
13. Cohn, D.: Measure Theory. Birkhäuser, Boston (1980)
14. Fuhrman, M., Tessitore, G.: Nonlinear Kolmogorov equations in infinite dimensional spaces: the backward stochastic differential equations approach and applications to optimal control. Ann. Probab. **30**(3), 1397–1465 (2002)
15. Grecksch, W., Tudor, C.: Stochastic evolution equations. A Hilbert Space Approach. Mathematical Research, vol. 85. Akademie-Verlag, Berlin (1995)
16. Gyöngy, I., Krylov, N.V.: On stochastics equations with respect to semimartingales. I. Stochastics **4**(1), 1–21 (1980)
17. Hu, Y., Peng, S.: Maximum principle for semilinear stochastic evolution control systems. Stochastics **33**(3–4), 159–180 (1990)
18. Hu, Y., Peng, S.: Adapted solution of a backward semilinear stochastic evolution equation. Stoch. Anal. Appl. **9**(4), 445–459 (1991)
19. Hu, Y., Peng, S.: Maximum principle for optimal control of stochastic system of functional type. Stoch. Anal. Appl. **14**(3), 283–301 (1996)
20. Imkeller, P., Reveillac, A., Richter, A.: Differentiability of quadratic BSDEs generated by continuous martingales. Ann. Appl. Probab. **22**(1), 285–336 (2012)
21. Krylov, N.V., Rozovskii, B.: Stochastic evolution equations. Stochastic Differential Equations: Theory and Applications. Interdisciplinary Mathematical Sciences, vol. 2, pp. 1–69. World Science Publisher, Hackensack (2007)

22. Li, X.J., Yong, J.M.: Optimal Control Theory for Infinite-Dimensional Systems. Systems & Control: Foundations & Applications. Birkhauser Boston Inc., Boston (1995)
23. Morlais, M.-A.: Quadratic BSDEs driven by a continuous martingale and applications to the utility maximization problem. Finance Stoch. **13**(1), 121–150 (2009)
24. Mazliak, L.: The maximum principle in stochastic control and backward equations. Backward Stochastic Differential Equations, Paris, 1995–1996. Pitman Research Notes in Mathematics Series, vol. 364, pp. 101–113. Longman, Harlow (1997)
25. Métivier, M., Pellaumail, J.: Stochastic Integration. Probability and Mathematical Statistics. Academic Press, Harcourt Brace Jovanovich, Publishers, New York (1980)
26. Øksendal, B., Proske, F., Zhang, T.: Backward stochastic partial differential equations with jumps and application to optimal control of random jump fields. Stochastics **77**(5), 381–399 (2005)
27. Pardoux, É., Peng, S.: Adapted solution of a backward stochastic differential equation. Syst. Control Lett. **14**(1), 55–61 (1990)
28. Peng, S.: A general stochastic maximum principle for optimal control problems. SIAM J. Control Optim. **28**(4), 966–979 (1990)
29. Peng, S.: Backward stochastic differential equations and applications to optimal control. Appl. Math. Optim. **27**(2), 125–144 (1993)
30. Pontryagin, L.S.: Optimal regulation processes. Amer. Math. Soc. Transl. **18**(2), 321–339 (1961)
31. Revuz, D., Yor, M.: Continuous Martingales and Brownian Motion. Grundlehren der Mathematischen Wissenschaften, Fundamental Principles of Mathematical Sciences, vol. 293, 3rd edn. Springer, Berlin (1999)
32. Rozovskiĭ, B.L.: Stochastic evolution systems. Linear Theory and Applications to Nonlinear Filtering, Translated from the Russian by A. Yarkho. Mathematics and its Applications, Soviet Series, vol. 35. Kluwer Academic Publishers Group, Dordrecht (1990)
33. Tang, S., Li, X.: Maximum principle for optimal control of distributed parameter stochastic systems with random jumps. Differential Equations, Dynamical Systems, and Control Science. notes in pure and applied mathematics, vol. 152, pp. 867–890. Dekker, New York (1994)
34. Tudor, C.: Optimal control for semilinear stochastic evolution equations. Appl. Math. Optim. **20**(3), 319–331 (1989)
35. Xu, W.S.: Stochastic maximum principle for optimal control problem of forward and backward system. J. Aust. Math. Soc. Ser. B **37**(2), 172–185 (1995)
36. Yong, J., Zhou, X.Y.: Stochastic Controls. Hamiltonian Systems and HJB Equations. Springer, New York (1999)
37. Zhou, X.Y.: On the necessary conditions of optimal controls for stochastic partial differential equations. SIAM J. Control Optim. **31**(6), 1462–1478 (1993)

Sufficient Conditions of Optimality for Forward-Backward Doubly SDEs with Jumps

AbdulRahman Al-Hussein and Boulakhras Gherbal

Abstract We consider a sufficient maximum principle of optimal control for a stochastic control problem. This problem is governed by a system of fully coupled multi-dimensional forward-backward doubly stochastic differential equation with Poisson jumps. Moreover, all the coefficients appearing in this system are allowed to be random and depend on the control variable. We derive, in particular, sufficient conditions for optimality for this stochastic optimal control problem. We apply our result to treat a kind of forward-backward doubly stochastic linear quadratic optimal control problems with jumps.

Keywords Poisson process · Sufficient conditions of optimality · Optimal control · Forward-backward doubly stochastic differential equation · Adjoint equations

Mathematical Subject Classification 2010 60H10 · 93E20 · 60G55

1 Introduction

Forward-backward stochastic differential equations (FBSDEs in short) were first studied by Antonelli in [2], and since then they are encountered in stochastic optimal control problem, which is one of the central themes of modern control science. For example, Xu in [20] studied a non-coupled continuous forward-backward stochastic control system. Then Wu [16], studied extensively the maximum principle for optimal control problem of fully coupled continuous forward-backward stochastic system. We refer the reader also to [3]. Peng and Wu [9], considered fully coupled continuous

A. Al-Hussein (✉)
Department of Mathematics, College of Science, Qassim University, P.O. Box 6644, Buraydah 51452, Saudi Arabia
e-mail: alhusseinqu@hotmail.com; hsien@qu.edu.sa

B. Gherbal
Laboratory of Applied Mathematics, University of Mohamed Khider, P.O. Box 145, Biskra 07000, Algeria
e-mail: bgherbal@yahoo.fr

M. Eddahbi et al. (eds.), *Statistical Methods and Applications in Insurance and Finance*, Springer Proceedings in Mathematics & Statistics 158,
DOI 10.1007/978-3-319-30417-5_7

forward-backward stochastic differential equations with random coefficients and applications to optimal control. A method of continuation is developed there. In this respect one can see also [22]. Shi and Wu in [11] studied the maximum principle for fully coupled continuous forward-backward stochastic system and provided under non-convexity assumption on the control domain necessary optimality conditions. The forward diffusion there does not contain the control variable.

Fully coupled FBSDEs with respect to Brownian motion and Poisson process were considered by Wu in [17] and Yin and Situ in [21]. Such equations have been shown to be very useful for example in studying linear quadratic optimal control problems of random jumps, and also to handle nonzero-sum differential games with random jumps. The work of Wu and Wang in [18] is useful in this respect. In [6], Øksendal and Sulem investigated stochastic maximum principle for non-coupled one-dimensional FBSDEs with jumps. In [13], Shi and Wu obtained both necessary and sufficient maximum principles for optimal control of FBSDEs with random jumps, when the control domain is convex. The result is applied to a mean-variance portfolio selection mixed with a recursive utility functional optimization problem. Meng [4], considered an optimal control problem of fully coupled forward-backward stochastic systems with Poisson jumps under partial information. More generally, Shi in [10] provided recently necessary conditions for optimal control of fully coupled FBSDEs with random jumps.

Backward doubly stochastic differential equations were first introduced by Pardoux and Peng in [7]. They gave a probabilistic representation of quasi linear stochastic partial differential equations. This important paper has given rise to a huge literature on BDSDEs and has become a powerful tool in many fields such as financial mathematics, optimal control, stochastic games, quasi linear partial differential equations. In 2003, Peng and Shi [8], introduced fully coupled forward-backward doubly stochastic differential equations (FBDSDEs in short). Such equations are generalizations of stochastic Hamilton systems. Existence and uniqueness of the solutions to (continuous) FBDSDEs with arbitrarily fixed time duration and under some monotone assumptions are established. Peng and Shi provided in [8] a probabilistic interpretation for the solutions of a class of quasilinear SPDEs. Moreover, in this respect, we refer the reader to [24] for an application of fully coupled FBDSDEs to provide a probabilistic formula for the solution of a quasilinear stochastic partial differential-integral equation (SPDIE in short). Another application to SPDEs can be found in [23]. These are some examples to show the importance and motivations to study FBDSDEs.

The existence and uniqueness of measurable solutions to FBDSDEs with Poisson jumps are established in [24] via the method of continuation. We refer the reader also to [8, 25] in this respect.

A sufficient maximum principle with partial information for a one-dimensional FBDSDE with jump with a forward equation being independent of the processes of the backward equation was studied in [19]. Necessary optimality conditions for FBDSDEs in [19] were derived also there under non-convexity assumption on the control domain. On the other hand, in [23], Zhang and Shi studied the maximum principle to find necessary conditions and sufficient conditions for optimality for a stochastic control problem governed by a continuous FBDSDE in dimension one. They allow also all the coefficients of these equations to contain control variables.

The general case of deriving the maximum principle for control problems governed by a multi-dimensional discontinuous FBDSDE with its coefficients being allowed to be random and depend on the control variable and when the control domain is not convex is still an interesting incomplete research problem. In the present work we shall consider this discontinuous situation, and study, in particular, a stochastic control problem where the system is governed by a nonlinear fully coupled multi-dimensional FBDSDE with jumps as in system (1) below. More precisely, we shall allow both the forward and backward equations to have random jumps, and establish sufficient conditions for optimality in the form of the maximum principle with a convex control domain. We will allow also all the coefficients appearing in our system to be random and contain control variables. Our results here are new in this respect. We will consider some relevant necessary optimality conditions for this problem (i.e. when we have a non-convex control domain) in a future work.

Our system under study is the following:

$$\begin{cases} dy_t = b(t, y_t, Y_t, z_t, Z_t, k_t, v_t)dt + \sigma(t, y_t, Y_t, z_t, Z_t, k_t, v_t)dW_t \\ \qquad\qquad + \int_\Theta \varphi(t, y_t, Y_t, z_t, Z_t, k_t, v_t, \rho)\tilde{N}(d\rho, dt) - z_t\overleftarrow{dB}_t, \\ dY_t = -f(t, y_t, Y_t, z_t, Z_t, k_t, v_t)dt - g(t, y_t, Y_t, z_t, Z_t, k_t, v_t)\overleftarrow{dB}_t \\ \qquad\qquad + Z_t dW_t + \int_\Theta k_t(\rho)\tilde{N}(d\rho, dt), \\ y_0 = x_0, Y_T = h(y_T), \end{cases} \tag{1}$$

where b, σ, φ, f and g are given mappings, with properties to be mentioned clearly in Sect. 2, and h is a combination of a square integrable random variable in $\mathbb{R}^m$ and a linear function from $\mathbb{R}^n$ to $\mathbb{R}^m$. The processes $(W_t)_{t\geq 0}$ and $(B_t)_{t\geq 0}$ are independent Brownian motions taking their values respectively in $\mathbb{R}^d$ and $\mathbb{R}^l$, while $v_.$ represents a control process, and $\tilde{N}(d\rho, dt)$ is the compensated Poisson random measure associated with a Poisson point process η. Here T is a fixed positive number.

We shall be interested in minimizing the cost functional

$$J(v_.) = \mathbb{E}\Big[\int_0^T \ell(t, y_t, Y_t, z_t, Z_t, k_t, v_t)dt + \beta(y_T) + \gamma(Y_0)\Big], \tag{2}$$

over the set of all admissible controls (to be described in Sect. 2 below).

Our formulation of these equations as well as cost functionals are given in abstract forms to allow the possibility to cover most of the applications available in the literature. For instance, a linear quadratic case can be given as a concrete and useful example. For more details of this example, we refer the reader to [12] or [18]. In fact, many applications of FBDSDE either in finance or to SPDEs can be developed in parallel to those provided in the literature; see e.g. [8, 23, 24]. We point out in particular that fully coupled FBDSDEs (1) provide a probabilistic formula for the solution of a quasilinear SPDIE in the sense of [23], and so one can apply our results here to study some optimal control problems of SPDIE (again similarly to [23]).

On the other hand, recently many research attentions have been drawn towards studying optimal control problems for discontinuous stochastic systems, especially those having Poisson jumps. One can see [1, 4, 6, 10, 17–19, 21, 24]. Many empirical studies have proven the existence of jumps in stock, foreign exchange and bond markets. There is, in fact, a trustworthy evidence that the dynamics of prices of financial instruments exhibit jumps that cannot be adequately captured solely by only diffusion processes. More precisely, jumps constitute a central feature in describing credit risk sensitive instruments.

Other models with jumps have also become popular in other areas of science and technology. As a result, paying more attention to study more applications of such discontinuous systems, including ours here is in demand. This will be somehow our coming future work.

Let us now close the introduction by recording that, in the case of partial information, one could easily develop similar results. We refer the reader to Ahmed et al. [1], in this respect.

We shall organize the paper as follows. In Sect. 2, we formulate the problem and give various assumptions used throughout the paper. In Sect. 3 we introduce the adjoint equation of (1), state our main theorem and give an example to illustrate this theorem. Section 4 is devoted to proving the main result.

2 Formulation of the Problem and Assumptions

Let $(\Omega, \mathscr{F}, \mathbb{P})$ be a complete probability space. Let $(W_t)_{t\in[0,T]}$ and $(B_t)_{t\in[0,T]}$ be two Brownian motions taking their values in $\mathbb{R}^d$ and $\mathbb{R}^l$ respectively. Let η be a Poisson point process taking its values in a measurable space $(\Theta, \mathscr{B}(\Theta))$. We denote by $\nu(d\rho)$ the characteristic measure of η which is assumed to be a σ-finite measure on $(\Theta, \mathscr{B}(\Theta))$, by $N(d\rho, dt)$ the Poisson counting measure (jump measure) induced by η with compensator $\nu(d\rho)dt$, and by

$$\tilde{N}(d\rho, dt) = N(d\rho, dt) - \nu(d\rho)dt,$$

the compensation of the jump measure $N(\cdot,\cdot)$ of η. Hence $\nu(O) = \mathbb{E}[N(O, 1)]$ for $O \in \mathscr{B}(\Theta)$. We assume that these three processes W, B and η are mutually independent.

Let $\mathscr{N}$ denote the class of $\mathbb{P}$-null sets of $\mathscr{F}$. For each $t \in [0, T]$, we define $\mathscr{F}_t := \mathscr{F}_t^W \vee \mathscr{F}_{t,T}^B \vee \mathscr{F}_t^\eta$, where for any process $\{\pi_t\}$, we set

$$\mathscr{F}_{s,t}^\pi = \sigma(\pi_r - \pi_s;\, s \leq r \leq t) \vee \mathscr{N},\ \mathscr{F}_t^\pi = \mathscr{F}_{0,t}^\pi.$$

For a Euclidean space E, let $\mathscr{M}^2(0, T; E)$ denote the set of jointly measurable processes $\{X_t, t \in [0, T]\}$ taking values in E, and satisfy: X_t is $\mathscr{F}_t$-measurable for a.e. $t \in [0, T]$, and

$$\mathbb{E}\Big[\int_0^T |X_t|_E^2\, dt\Big] < \infty.$$

Let $L_\nu^2(E)$ be the set of $\mathscr{B}(\Theta)$-measurable mapping k with values in E such that

$$|||k||| := \Big[\int_\Theta |k(\rho)|_E^2\, \nu(d\rho)\Big]^{\frac{1}{2}} < \infty.$$

Denote by $\mathcal{N}_\eta^2(0, T; E)$ to the set of processes $\{K_t,\ t \in [0, T]\}$ that take their values in $L_\nu^2(E)$ and satisfy: K_t is $\mathscr{F}_t$-measurable for a.e. $t \in [0, T]$, and

$$\mathbb{E}\Big[\int_0^T \int_\Theta |K_t(\rho)|_E^2\, \nu(d\rho)dt\Big] < \infty.$$

Finally, we set

$$\begin{aligned}\mathbb{M}^2 := \mathcal{M}^2\left(0, T; \mathbb{R}^n\right) \times \mathcal{M}^2\left(0, T; \mathbb{R}^m\right) \times \mathcal{M}^2\left(0, T; \mathbb{R}^{n\times l}\right)\\ \times \mathcal{M}^2\left(0, T; \mathbb{R}^{m\times d}\right) \times \mathcal{N}_\eta^2\left(0, T; \mathbb{R}^m\right).\end{aligned}$$

Then $\mathbb{M}^2$ is a Hilbert space with respect to the norm $\|\cdot\|_{\mathbb{M}^2}$ given by

$$\|\zeta.\|_{\mathbb{M}^2}^2 := \mathbb{E}\Big[\int_0^T \left(|y_t|^2 + |Y_t|^2 + \|z_t\|^2 + \|Z_t\|^2 + |||k_t|||^2\right)dt\Big],$$

for $\zeta. = (y., Y., z., Z., k.)$.

Let U be a non-empty convex subset of $\mathbb{R}^r$. We say that $v. : [0, T] \times \Omega \to \mathbb{R}^r$ is *admissible* if $v. \in \mathcal{M}^2(0, T; \mathbb{R}^r)$ and $v_t \in U\ \ a.e.,\ \mathbb{P} - a.s.$ The set of admissible controls will be denoted by $\mathscr{U}_{ad}$. Consider the following controlled fully coupled FBDSDE with jumps:

$$\begin{cases} dy_t = b(t, y_t, Y_t, z_t, Z_t, k_t, v_t)dt + \sigma(t, y_t, Y_t, z_t, Z_t, k_t, v_t)dW_t \\ \qquad + \int_\Theta \varphi(t, y_t, Y_t, z_t, Z_t, k_t, v_t, \rho)\tilde{N}(d\rho, dt) - z_t\overleftarrow{dB}_t, \\ dY_t = -f(t, y_t, Y_t, z_t, Z_t, k_t, v_t)dt - g(t, y_t, Y_t, z_t, Z_t, k_t, v_t)\overleftarrow{dB}_t \\ \qquad + Z_t dW_t + \int_\Theta k_t(\rho)\tilde{N}(d\rho, dt), \\ y_0 = x_0,\ Y_T = h(y_T), \end{cases} \tag{3}$$

where the mappings

$$\begin{aligned} &b : \Omega \times [0, T] \times \mathbb{R}^n \times \mathbb{R}^m \times \mathbb{R}^{n\times l} \times \mathbb{R}^{m\times d} \times L_\nu^2(\mathbb{R}^m) \times \mathbb{R}^r \to \mathbb{R}^n, \\ &\sigma : \Omega \times [0, T] \times \mathbb{R}^n \times \mathbb{R}^m \times \mathbb{R}^{n\times l} \times \mathbb{R} \times L_\nu^2(\mathbb{R}^m) \times \mathbb{R}^r \to \mathbb{R}^{n\times d}, \\ &\varphi : \Omega \times [0, T] \times \mathbb{R}^n \times \mathbb{R}^m \times \mathbb{R}^{n\times l} \times \mathbb{R}^{m\times d} \times L_\nu^2(\mathbb{R}^m) \times \mathbb{R}^r \times \Theta \to \mathbb{R}^n, \\ &f : \Omega \times [0, T] \times \mathbb{R}^n \times \mathbb{R}^m \times \mathbb{R}^{n\times l} \times \mathbb{R}^{m\times d} \times L_\nu^2(\mathbb{R}^m) \times \mathbb{R}^r \to \mathbb{R}^m, \end{aligned}$$

$$g : \Omega \times [0, T] \times \mathbb{R}^n \times \mathbb{R}^m \times \mathbb{R}^{n\times l} \times \mathbb{R} \times L^2_\nu(\mathbb{R}^m) \times \mathbb{R}^r \to \mathbb{R}^{m\times l},$$
$$h : \Omega \times \mathbb{R}^n \to \mathbb{R}^m,$$

are measurable (further properties to be introduced later in this section) and $v_\cdot \in \mathcal{U}_{ad}$. Given a full-rank $m \times n$ matrix R of real indices, we assume that h is defined, for $(\omega, x) \in \Omega \times \mathbb{R}^n$, by $h(\omega, x) := c\, Rx + \xi(\omega)$, where $c \neq 0$ is a constant and ξ is a fixed arbitrary element of $L^2(\Omega, \mathcal{F}_T, \mathbb{P}; \mathbb{R}^m)$. This formula of h is useful and vital in establishing the proof our main theorem (Theorem 3) in Sect. 4, when we apply in particular the adjoint equations of (3).

Note that the integral with respect to $\overleftarrow{dB}$ is a "backward" Itô integral, while the integral with respect to dW is a standard "forward" Itô integral. We refer the reader to [5] for more details on such integrals, which are particular cases of the Itô-Skorohod stochastic integral.

A solution of (3) is a quintuple (y, Y, z, Z, k) of stochastic processes such that (y, Y, z, Z, k) belongs to $\mathbb{M}^2$ and satisfies the following FBDSDE:

$$\begin{cases} y_t = x_0 + \int_0^t b(s, y_s, Y_s, z_s, Z_s, k_s, v_s)ds + \int_0^t \sigma(s, y_s, Y_s, z_s, Z_s, k_s, v_s)dW_s \\ \qquad + \int_0^t \int_\Theta \varphi(s, y_s, Y_s, z_s, Z_s, k_s, v_s, \rho)\tilde{N}(d\rho, ds) - \int_0^t z_s \overleftarrow{dB}_s, \\ \\ Y(t) = h(y_T) + \int_t^T f(s, y_s, Y_s, z_s, Z_s, k_s, v_s)ds + \int_t^T g(s, y_s, Y_s, z_s, Z_s, k_s, v_s)\overleftarrow{dB}_s \\ \qquad - \int_t^T Z_s dW_s - \int_t^T \int_\Theta k_s(\rho)\tilde{N}(d\rho, ds), \ t \in [0, T]. \end{cases}$$

Define the cost functional by:

$$J(v_\cdot) := \mathbb{E}\Big[\int_0^T \ell(t, y_t, Y_t, z_t, Z_t, k_t, v_t)dt + \beta(y_T) + \gamma(Y_0)\Big], \ v_\cdot \in \mathcal{U}_{ad}, \tag{4}$$

where

$$\beta : \mathbb{R}^n \to \mathbb{R},$$
$$\gamma : \mathbb{R}^m \to \mathbb{R},$$
$$\ell : \Omega \times [0, T] \times \mathbb{R}^n \times \mathbb{R}^m \times \mathbb{R}^{n\times l} \times \mathbb{R}^{m\times d} \times L^2_\nu(\mathbb{R}^m) \times \mathbb{R}^r \to \mathbb{R}$$

are measurable functions such that (4) is defined.

Now the control problem of system (3) is to minimize J over $\mathcal{U}_{ad}$. In this case we say that $u_\cdot \in \mathcal{U}_{ad}$ is an *optimal control* if

$$J(u_\cdot) = \inf_{v_\cdot \in \mathcal{U}_{ad}} J(v_\cdot). \tag{5}$$

Let us set the following notations:

$$\zeta = (y, Y, z, Z, k) \in \mathbb{R}^{n+m+n\times l+m\times d} \times L^2_\nu(\mathbb{R}^m),$$
$$A(t, \zeta, v) = (-R^* f, Rb, -R^* g, R\sigma, R\varphi)(t, \zeta, v),$$
$$\langle A, \zeta\rangle = -\big\langle y, R^* f\big\rangle + \langle Y, Rb\rangle - \big\langle z, R^* g\big\rangle + \langle Z, R\sigma\rangle + \langle\langle k, R\varphi\rangle\rangle,$$

where

$$R^*g = (R^*g_1 \cdots R^*g_l),\ R\sigma = (R\sigma_1 \cdots R\sigma_d), \ldots,$$

by using the columns $\{g_1, \ldots, g_l\}$ and $\{\sigma_1, \ldots, \sigma_d\}$ of g and σ respectively, and

$$\langle\langle k, R\varphi\rangle\rangle\,(t, \zeta, v) = \int_\Theta \langle k(\rho), R\varphi(t, \zeta, v, \rho)\rangle\, \nu(d\rho).$$

The following assumptions will be our main assumptions in the paper. We shall mimic similar assumptions from the literature (e.g. [23]) for this purpose.

- (A1) $\forall \zeta = (y, Y, z, Z, k), \bar{\zeta} = (\bar{y}, \bar{Y}, \bar{z}, \bar{Z}, \bar{k}) \in \mathbb{R}^{n+m+n\times l+m\times d} \times L^2_\nu(\mathbb{R}^m)$, $\forall t \in [0, T],\ \forall v \in \mathbb{R}^r$,

$$\begin{aligned}\left\langle A(t, \zeta, v) - A(t, \bar{\zeta}, v), \zeta - \bar{\zeta}\right\rangle \ &\le -\mu_1(|R(y - \bar{y})|^2 + \left|R^*(Y - \bar{Y})\right|^2) \\ &\quad -\mu_2(\|R(z - \bar{z})\|^2 + \left\|R^*(Z - \bar{Z})\right\|^2 + \left|\left\|R^*(k - \bar{k})\right\|\right|^2),\end{aligned}$$

and

$$c > 0,$$

or
- (A1)'

$$\begin{aligned}\left\langle A(t, \zeta, v) - A(t, \bar{\zeta}, v), \zeta - \bar{\zeta}\right\rangle \ &\ge \mu_1(|R(y - \bar{y})|^2 + \left|R^*(Y - \bar{Y})\right|^2) \\ &\quad +\mu_2(\|R(z - \bar{z})\|^2 + \left\|R^*(Z - \bar{Z})\right\|^2 + \left|\left\|R^*(k - \bar{k})\right\|\right|^2),\end{aligned}$$

and

$$c < 0,$$

where μ_1 and μ_2 are nonnegative constants with $\mu_1 + \mu_2 > 0$. Moreover $\mu_1 > 0$ (resp. $\mu_2 > 0$) when $m > n$ (resp. $n > m$).
- (A2) For each $\zeta \in \mathbb{R}^{n+m+n\times l+m\times d} \times L^2_\nu(\mathbb{R}^m), v \in \mathbb{R}^r,\ A(t, \zeta, v) \in \mathbb{M}^2$.
- (A3) We assume that

$$\begin{cases} (i) & \text{the mappings } f, b, g, \sigma, \varphi, \ell \text{ are continuously differentiable with respect to } (y, Y, z, Z, k, v), \text{ and } \beta \text{ and } \gamma \text{ are continuously differentiable with respect to } y \text{ and } Y, \text{ respectively,} \\ (ii) & \text{the derivatives of } f, b, g, \sigma, \varphi \text{ with respect to the above arguments are bounded,} \\ (iii) & \text{the derivatives of } \ell \text{are bounded by } C(1 + |y| + |Y| + \|z\| + \|Z\| + |\|k\||), \\ (iv) & \text{the derivatives of } \beta \text{ and } \gamma \text{ are bounded by } C\,(1 + |y|) \text{ and } C\,(1 + |Y|) \text{ respectively,} \end{cases}$$

for some constant $C > 0$.

Remark 1 The condition $c > 0$ in (A1) guarantees the following monotonicity condition of the mapping h:

$$\langle h(y) - h(\bar{y}), R(y - \bar{y})\rangle \geq c\,|R(y - \bar{y})|^2, \quad \forall\, y, \bar{y} \in \mathbb{R}^n.$$

The same thing happens also for $c < 0$ in (A1)'.

The following theorem is concerned with the existence and uniqueness of the solution of (3).

Theorem 1 *For any given admissible control* $v_{\cdot}$, *if assumptions (A1)–(A3) (or (A1)', (A2), (A3)) hold, then* (3) *has a unique solution.*

Our assumptions in this theorem satisfy the assumptions of the corresponding result in [24], so the proof of this theorem can be gleaned from [24]. We refer the reader also to [8, 25] for useful works in this respect.

3 Adjoint Equations and the Maximum Principle

Suppose that (A1)–(A3) hold. We want to introduce the adjoint equations of FBDSDE (3) and then present our main result of the maximum principle for our optimal control problem governed by the FBDSDE with jumps (3). To this end, let us begin by defining the Hamiltonian H from $[0, T] \times \Omega \times \mathbb{R}^n \times \mathbb{R}^m \times \mathbb{R}^{n\times l} \times \mathbb{R}^{m\times d} \times L^2_\nu(\mathbb{R}^m) \times \mathbb{R}^r \times \mathbb{R}^m \times \mathbb{R}^n \times \mathbb{R}^{m\times l} \times \mathbb{R}^{n\times d} \times L^2_\nu(\mathbb{R}^n)$ to $\mathbb{R}$ by the formula:

$$\begin{aligned}
&H(t, y, Y, z, Z, k, v, p, P, q, Q, V) :\\
&\quad= \langle p, f(t, y, Y, z, Z, k, v)\rangle\\
&\qquad- \langle P, b(t, y, Y, z, Z, k, v)\rangle + \langle q, g(t, y, Y, z, Z, k, v)\rangle\\
&\qquad- \langle Q, \sigma(t, y, Y, z, Z, k, v)\rangle - \ell(t, y, Y, z, Z, k, v)\\
&\qquad- \int_\Theta \langle V(\rho), \varphi(t, y, Y, z, Z, k, v, \rho)\rangle\, \nu(d\rho).
\end{aligned} \tag{6}$$

Let $v_{\cdot}$ be an arbitrary element of $\mathscr{U}_{ad}$ and $\{(y_t, Y_t, z_t, Z_t, k_t),\ t \in [0, T]\}$ be the corresponding solution of (3). The adjoint equations of our FBDSDE with jumps (3) are

$$\begin{cases}
dp_t = H_Y(t)dt + H_Z(t)dW_t - q_t\overleftarrow{dB}_t + \int_\Theta H_k(t)\tilde{N}(d\rho, dt),\\
dP_t = H_y(t)dt + H_z(t)\overleftarrow{dB}_t + Q_t dW_t + \int_\Theta V_t(\rho)\tilde{N}(d\rho, dt),\\
p_0 = -\gamma_Y(Y_0),\ P_T = -c\,R^*\,p_T + \beta_y(y_T),
\end{cases} \tag{7}$$

where and $\beta_y(y_T)$ is the gradient $\nabla_y\beta(y_T) \in \mathbb{R}^n$ and $H_y(t)$ is the gradient $\nabla_y H(t, y, Y_t, z_t, Z_t, k_t, v_t, p_t, P_t, q_t, Q_t, V_t) \in \mathbb{R}^n, \ldots$ etc. Let us in the following say some thing more about this system (7).

Theorem 2 *Under (A1)–(A3) there exists a unique solution* (p, P, q, Q, V) *of the adjoint equations* (7) *(in* $\tilde{\mathbb{M}}^2 := \mathscr{M}^2(0,T;\mathbb{R}^m) \times \mathscr{M}^2(0,T;\mathbb{R}^n) \times \mathscr{M}^2\left(0,T;\mathbb{R}^{m\times l}\right) \times \mathscr{M}^2\left(0,T;\mathbb{R}^{n\times d}\right) \times \mathscr{N}_\eta^2(0,T;\mathbb{R}^n)$*).*

Proof This system (7) can be rewritten as in the following system:

$$\begin{cases} dp_t = (f_Y^*(t)p_t - b_Y^*(t)P_t + g_Y^*(t)q_t - \sigma_Y^*(t)Q_t - \int_\Theta \varphi_Y^*(t)V_t(\rho)\nu(d\rho) - \ell_Y(t))dt \\ \qquad + (f_Z^*(t)p_t - b_Z^*(t)P_t + g_Z^*(t)q_t - \sigma_Z^*(t)Q_t - \int_\Theta \varphi_Z^*(t)V_t(\rho)\nu(d\rho) - \ell_Z(t))dW_t \\ \qquad - q_t\overleftarrow{dB}_t + \int_\Theta (f_k^*(t)p_t - b_k^*(t)P_t + g_k^*(t)q_t - \sigma_k^*(t)Q_t \\ \qquad - (\varphi_k^*(t)V_t)(\rho) - \ell_k(t))\tilde{N}(d\rho,dt), \\ \\ dP_t = (f_y^*(t)p_t - b_y^*(t)P_t + g_y^*(t)q_t - \sigma_y^*(t)Q_t - \int_\Theta \varphi_y^*(t)V_t(\rho)\nu(d\rho) - \ell_y(t))dt \\ \qquad + (f_z^*(t)p_t - b_z^*(t)P_t + g_z^*(t)q_t - \sigma_z^*(t)Q_t - \int_\Theta \varphi_z^*(t)V_t(\rho)\nu(d\rho) - \ell_z(t))\overleftarrow{dB}_t \\ \qquad + Q_t dW_t + \int_\Theta V_t(\rho)\tilde{N}(d\rho,dt), \\ \\ p_0 = -\gamma_Y(Y_0),\ P_T = -c\,R^*(t)\,p_T + \beta_y(y_T), \end{cases}$$

which is a linear FBDSDE with jumps. Here $f_y^*(t) \in \mathbb{R}^{n\times m}$ is the adjoint (transpose) of the Fréchet derivative (hence the Gâteaux derivative) $D_y f(t,y,Y_t,z_t,Z_t,k_t,v_t) \in \mathbb{R}^{m\times n}$ of $f(t,\cdot,Y_t,z_t,Z_t,k_t,v_t) : \mathbb{R}^n \to \mathbb{R}^m$ at $y, \ldots$ etc. Thanks to assumptions (A1)–(A3) this latter linear FBDSDE satisfy easily (A1)', (A2) and (A3). In fact the monotonicity condition follows from the definition of Gâteaux derivatives (as limits) and the fact that the corresponding mappings satisfy originally $R-$monotonicity condition in (A1). Thus the desired result follows from Theorem 1.

Now our main theorem is the following.

Theorem 3 *Assume that (A1)–(A3) hold. Given* $u_\cdot \in \mathscr{U}_{ad}$*, let* (y, Y, z, Z, k) *and* (p, P, q, Q, V) *be the corresponding solutions of the FBDSDEs* (3) *and* (7) *respectively. Suppose that the following assumptions hold*:
(i) β *and* γ *are convex.*
(ii) For all $t \in [0,T]$, $\mathbb{P}$*-a.s., the function* $H(t,\cdot,\cdot,\cdot,\cdot,\cdot,\cdot,p_t,P_t,q_t,Q_t,V_t)$ *is concave.*
(iii) We have

$$\begin{aligned} &H(t,y_t,Y_t,z_t,Z_t,k_t,u_t,p_t,P_t,q_t,Q_t,V_t) \\ &\qquad = \max_{v\in U} H(t,y_t,Y_t,z_t,Z_t,k_t,v,p_t,P_t,q_t,Q_t,V_t), \end{aligned} \tag{8}$$

for a.e, $\mathbb{P}$*-a.s.*

Then $(y, Y, z, Z, k, u_\cdot)$ *is an optimal solution of the control problem* (3)–(5).

Remark 2 Condition (A1) assumed in this theorem and the lemmas that follow is only needed to guarantee the existence and uniqueness of the solutions of (3) and (7), and so if we can get these unique solutions without assuming (A1) there will not any necessity to assume it in advance in this theorem.

The proof of Theorem 3 will be established in Sect. 4. Now to illustrate this theorem let us present an example.

Example 1 Let $(\Theta, \mathscr{B}(\Theta)) = ([0,1], \mathscr{B}([0,1]))$. Let $\tilde{N}(d\rho, dt)$ be a compensated Poisson random measure, where $(t, \rho) \in [0,1] \times [0,1]$. Recall that $\mathbb{E}[\tilde{N}(d\rho, dt)^2] = \nu(d\rho)dt$ is required to be a finite Borel measure such that $\int_{[0,1]} \rho^2 \nu(d\rho) < \infty$. Let the controls domain be $U = [-1, 1]$. Consider the following stochastic control system:

$$\begin{cases} dy_t = (1+t)v_t dt + (z_t - Z_t + v_t)dW_t - z_t \overleftarrow{dB}_t - \int_{[0,1]} (v_t \rho + k_t(\rho))\tilde{N}(d\rho, dt), \\ dY_t = -(t-4)v_t dt - \frac{3}{2}(z_t + Z_t + v_t)\overleftarrow{dB}_t + Z_t dW_t + \int_{[0,1]} k_t(\rho)\tilde{N}(d\rho, dt), \\ y_0 = Y_1 = x \in \mathbb{R},\ t \in (0,1), \end{cases} \tag{9}$$

where W, B are Brownian motions in $\mathbb{R}$, and W, B and $\tilde{N}$ are mutually independent. Consider also a cost functional given for $v_\cdot \in \mathscr{U}_{ad}$ by

$$J(v_\cdot) = \frac{1}{2}\mathbb{E}\Big[\int_0^1 (y_t^2 + Y_t^2 + z_t^2 + Z_t^2 + \int_{[0,1]} k_t^2(\rho)\nu(d\rho) + v_t^2)dt + y_1^2 + Y_0^2\Big], \tag{10}$$

with

$$\ell(t, y_t, Y_t, z_t, Z_t, k_t, v_t) = \frac{1}{2}(y_t^2 + Y_t^2 + z_t^2 + Z_t^2 + \int_{[0,1]} k_t^2(\rho)\nu(d\rho) + v_t^2),$$
$$\beta(y_t) = \frac{1}{2}y_t^2,\quad \gamma(Y_t) = \frac{1}{2}Y_t^2.$$

We define the value function by

$$J(u_\cdot) = \inf_{v_\cdot \in \mathscr{U}_{ad}} J(v_\cdot). \tag{11}$$

This system (9) can be related to the one in (3) by setting the following mappings:

$$\begin{aligned} b(t, y_t, Y_t, z_t, Z_t, k_t, v_t) &= (1+t)v_t, \\ \sigma(t, y_t, Y_t, z_t, Z_t, k_t, v_t) &= z_t - Z_t + v_t, \\ \varphi(t, y_t, Y_t, z_t, Z_t, k_t, v_t, \rho) &= -v_t\rho - k(\rho), \\ f(t, y_t, Y_t, z_t, Z_t, k_t, v_t) &= (t-4)v_t, \\ g(t, y_t, Y_t, z_t, Z_t, k_t, v_t) &= \frac{3}{2}(z_t + Z_t + v_t), \\ h(y_1) &= y_1,\ \text{i.e. } c = 1, \xi = 0. \end{aligned}$$

One can see easily that assumptions (A1), (A3) (and of course (A2)) are fulfilled for this system. More precisely, for (A1) we observe with the help of the notations preceding assumption (A1) and Cauchy-Schwartz inequality, that, for $\zeta = (y, Y, z, Z)$, $\bar{\zeta} = (\bar{y}, \bar{Y}, \bar{z}, \bar{Z}, \bar{k}) \in \mathbb{R} \times \mathbb{R} \times \mathbb{R} \times \mathbb{R} \times L_\nu^2(\mathbb{R})$,

$$\begin{aligned}\langle A(t,\zeta,v)-A(t,\bar{\zeta},v),\zeta-\bar{\zeta}\rangle &\leq -\frac{1}{2}||z-\bar{z}||^2-\frac{1}{2}(z-\bar{z})(Z-\bar{Z})\\&\quad-||Z-\bar{Z}||^2-|||k-\bar{k}|||^2\\&\leq -\frac{3}{4}\left(||z-\bar{z}||^2+||Z-\bar{Z}||^2+|||k-\bar{k}|||^2\right),\end{aligned}$$

which satisfies (A1) with $\mu_1=0$, $\mu_2=\frac{3}{4}$ and $c_1=1$.

Thus Theorem 1 guarantees the existence and uniqueness of the solution of (9). Note that (9)–(11) is a linear-quadratic optimal control problem with jumps.

Letting $u_\cdot\equiv 0$, we find from the construction of FBDSDEs with jumps (as for instance in [24]) that the corresponding solution (y_t,Y_t,z_t,Z_t,k_t) of (9) equals $(x,x,0,0,0)$, for all $t\in[0,1]$.

Next notice that the adjoint equations of (9) are

$$\begin{cases} dp_t=-Y_tdt+(\frac{3}{2}q_t+Q_t-Z_t)dW_t-q_t\overleftarrow{dB}_t-\int_{[0,1]}(V_t(\rho)+k_t(\rho))\tilde{N}(d\rho,dt),\\ dP_t=-y_tdt+(\frac{3}{2}q_t-Q_t-z_t)\overleftarrow{dB}_t+Q_tdW_t+\int_{[0,1]}V_t(\rho)\tilde{N}(d\rho,dt),\\ p_0=-x,\ P_1=-p_1+x,\ t\in(0,1).\end{cases} \tag{12}$$

Since p_0 is deterministic, then so is p_t. Hence

$$p_t=p_0-\int_0^t Y_tdt=p_0-x\int_0^t dt=-x(1+t).$$

Thus P_1 is deterministic since:

$$P_1=-p_1+x=3x.$$

It follows similarly that

$$P_t=P_1+\int_t^1 y_tdt=3x+x(1-t)=x(4-t).$$

In particular, $(p_t,P_t,q_t,Q_t,V_t)\equiv(-x(1+t),x(4-t),0,0,0)$ is the unique solution of (12). These facts show that the Hamiltonian attains an explicit formula:

$$\begin{aligned}&H(t,y_t,Y_t,z_t,Z_t,k_t,v,p_t,P_t,q_t,Q_t,V_t)\\&=p_t(t-4)v-P_t(1+t)v+\frac{3}{2}q_t(z_t+Z_t+v)-Q_t(z_t-Z_t+v)\\&+\int_{[0,1]}(v\rho+k_t(\rho))\,V_t(\rho)\,\nu(d\rho)-\frac{1}{2}\,(y_t^2+Y_t^2+z_t^2+Z_t^2\\&+\int_{[0,1]}k_t^2(\rho)\nu(d\rho)+v^2)\end{aligned}$$

$$= -x(1+t)(t-4)v - x(1+t)(4-t)v - \frac{1}{2}v^2 - \frac{1}{2}x^2 - \frac{1}{2}x^2$$
$$= -\frac{1}{2}v^2 - x^2, \ v \in U.$$

Hence

$$H(t, y_t, Y_t, z_t, Z_t, k_t, v, p_t, P_t, q_t, Q_t, V_t)$$
$$-H(t, y_t, Y_t, z_t, Z_t, k_t, u_t, p_t, P_t, q_t, Q_t, V_t)$$
$$= -\frac{1}{2}v^2 - x^2 + \frac{1}{2}u_t^2 + x^2 = -\frac{1}{2}v^2 \leq 0, \ \forall v \in U, \ a.e. \ t, \ \mathbb{P} - a.s.$$

As a result, condition (iii) of Theorem 3 holds here for $u_. = 0$. Furthermore, all other conditions of Theorem 3 can be verified easily. Consequently,

$$(y, Y, z, Z, k, u_.) \equiv (x, x, 0, 0, 0, 0)$$

is an optimal solution of the control problem (9)–(11).

For more applications of the theory of fully coupled FBDSDEs particularly in providing a probabilistic formula for the solution of a quasilinear SPDIE we refer the reader to [24, p. 15].

4 Proof of Theorem 3

In this section we shall establish the proof of Theorem 3. Let us recall first the following lemma.

Lemma 1 [*Integration by parts*] *Let* $(\alpha, \widehat{\alpha}) \in \left[\mathscr{S}^2(0, T; \mathbb{R}^n)\right]^2$, $(\beta, \widehat{\beta}) \in \left[\mathscr{M}^2(0, T; \mathbb{R}^n)\right]^2$, $(\gamma, \widehat{\gamma}) \in \left[\mathscr{M}^2(0, T; \mathbb{R}^{n\times k})\right]^2$, $(\delta, \widehat{\delta}) \in \left[\mathscr{M}^2(0, T; \mathbb{R}^{n\times d})\right]^2$, *and* $(K, \widehat{K}) \in \left[\mathscr{N}_\eta^2(0, T; \mathbb{R}^m)\right]^2$. *Assume that*

$$\alpha_t = \alpha_0 + \int_0^t \beta_s ds + \int_0^t \gamma_s \overleftarrow{dB}_s + \int_0^t \delta_s dW_s + \int_0^t \int_\Theta K_s(\rho) \tilde{N}(d\rho, ds),$$

and

$$\widehat{\alpha}_t = \widehat{\alpha}_0 + \int_0^t \widehat{\beta}_s ds + \int_0^t \widehat{\gamma}_s \overleftarrow{dB}_s + \int_0^t \widehat{\delta}_s dW_s + \int_0^t \int_\Theta \widehat{K}_s(\rho) \tilde{N}(d\rho, ds),$$

for $t \in [0, T]$. *Then*

$$\langle \alpha_T, \widehat{\alpha}_T \rangle = \langle \alpha_0, \widehat{\alpha}_0 \rangle + \int_0^T \langle \alpha_t, d\widehat{\alpha}_t \rangle + \int_0^T \langle \widehat{\alpha}_t, d\alpha_t \rangle + \int_0^T d \langle \alpha, \widehat{\alpha} \rangle_t .$$

$$\mathbb{E}[\langle \alpha_T, \widehat{\alpha}_T \rangle] = \mathbb{E}[\langle \alpha_0, \widehat{\alpha}_0 \rangle] + \mathbb{E}\Big[\int_0^T \langle \alpha_t, d\widehat{\alpha}_t \rangle\Big] + \mathbb{E}\Big[\int_0^T \langle \widehat{\alpha}_t, d\alpha_t \rangle\Big]$$
$$- \mathbb{E}\Big[\int_0^T \langle \gamma_t, \widehat{\gamma}_t \rangle dt\Big] + \mathbb{E}\Big[\int_0^T \langle \delta_t, \widehat{\delta}_t \rangle dt\Big] + \mathbb{E}\Big[\int_0^T \int_\Theta \langle K_t(\rho), \widehat{K}_t(\rho) \rangle \nu(d\rho) dt\Big].$$

This lemma can be deduced directly from Itô's formula with jumps (see e.g. [14, 15]).

We now prove Theorem 3. We start with two lemmas.

Lemma 2 *Assume (A1)–(A3). Let* $v.$ *be an arbitrary element of* $\mathscr{U}_{ad}$, *and let* $(y^{v.}, Y^{v.}, z^{v.}, Z^{v.}, k^{v.})$ *be the corresponding solution of* (3). *Then we have*

$$J(v.) - J(u.) \geq \mathbb{E}\big[\langle P_T, y_T^{v.} - y_T \rangle\big] + \mathbb{E}\big[\langle c R^* p_T, y_T^{v.} - y_T \rangle\big] - \mathbb{E}\big[\langle p_0, Y_0^{v.} - Y_0 \rangle\big]$$
$$+ \mathbb{E}\Big[\int_0^T \big(\ell(t, y_t^{v.}, Y_t^{v.}, z_t^{v.}, Z_t^{v.}, k_t^{v.}, v_t) - \ell(t, y_t, Y_t, z_t, Z_t, k_t, u_t)\big) dt\Big]. \tag{13}$$

Proof From (4) we get

$$J(v.) - J(u.) = \mathbb{E}\big[\beta(y_T^{v.}) - \beta(y_T)\big] + \mathbb{E}\big[\gamma(Y_0^{v.}) - \gamma(Y_0)\big]$$
$$+ \mathbb{E}\Big[\int_0^T \big(\ell(t, y_t^{v.}, Y_t^{v.}, z_t^{v.}, Z_t^{v.}, k_t^{v.}, v_t) - \ell(t, y_t, Y_t, z_t, Z_t, k_t, u_t)\big) dt\Big].$$

Since β and γ are convex, we obtain

$$\beta(y_T^{v.}) - \beta(y_T) \geq \langle \beta_y(y_T), y_T^{v.} - y_T \rangle,$$
$$\gamma(Y_0^{v.}) - \gamma(Y_0) \geq \langle \gamma_Y(Y_0), Y_0^{v.} - Y_0 \rangle,$$

which imply that

$$J(v.) - J(u.) \geq \mathbb{E}\big[\langle \beta_y(y_T), y_T^{v.} - y_T \rangle\big] + \mathbb{E}\big[\langle \gamma_Y(Y_0), Y_0^{v.} - Y_0 \rangle\big]$$
$$+ \mathbb{E}\Big[\int_0^T \big(\ell(t, y_t^{v.}, Y_t^{v.}, z_t^{v.}, Z_t^{v.}, k_t^{v.}, v_t) - \ell(t, y_t, Y_t, z_t, Z_t, k_t, u_t)\big) dt\Big].$$

But from the adjoint equation (6) and system (3) we know

$$p_0 = -\gamma_Y(Y_0), \; P_T = -c R^* p_T + \beta_y(y_T).$$

Thus (13) holds.

The following lemma contains duality relations between (3) and (7) (see the equivalent equations in the proof of Theorem 2).

Lemma 3 *Suppose that assumptions of Lemma* 2 (in particular (A1)–(A3)) hold. Then

$$
\begin{aligned}
&-\mathbb{E}\big[\big\langle p_0, Y_0^{v.} - Y_0\big\rangle\big] \\
&\quad = -\mathbb{E}\big[\big\langle p_T, Y_T^{v.} - Y_T\big\rangle\big] \\
&\qquad - \mathbb{E}\Big[\int_0^T \big\langle p_t, f(t, y_t^{v.}, Y_t^{v.}, z_t^{v.}, Z_t^{v.}, k_t^{v.}, v_t) - f(t, y_t, Y_t, z_t, Z_t, k_t, u_t)\big\rangle dt\Big] \\
&\qquad + \mathbb{E}\Big[\int_0^T \big\langle H_Y(t, y_t, Y_t, z_t, Z_t, k_t, u_t, p_t, P_t, q_t, Q_t, V_t), Y_t^{v.} - Y_t\big\rangle dt\Big] \\
&\qquad - \mathbb{E}\Big[\int_0^T \big\langle q_t, g(t, y_t^{v.}, Y_t^{v.}, z_t^{v.}, Z_t^{v.}, k_t^{v.}, v_t) - g(t, y_t, Y_t, z_t, Z_t, k_t, u_t)\big\rangle dt\Big] \\
&\qquad + \mathbb{E}\Big[\int_0^T \big\langle H_Z(t, y_t, Y_t, z_t, Z_t, k_t, u_t, p_t, P_t, q_t, Q_t, V_t), Z_t^{v.} - Z_t\big\rangle dt\Big] \\
&\qquad + \mathbb{E}\Big[\int_0^T \int_\Theta \langle H_k(t, y_t, Y_t, z_t, Z_t, k_t, u_t, p_t, P_t, q_t, Q_t, V_t), k_t^{v.}(\rho) \\
&\qquad\qquad - k_t(\rho)\rangle \nu(d\rho) dt\Big],
\end{aligned} \tag{14}
$$

and

$$
\begin{aligned}
&\mathbb{E}\big[\big\langle P_T, y_T^{v.} - y_T\big\rangle\big] \\
&\quad = \mathbb{E}\Big[\int_0^T \langle P_t, b(t, y_t^{v.}, Y_t^{v.}, z_t^{v.}, Z_t^{v.}, k_t^{v.}, v_t) - b(t, y_t, Y_t, z_t, Z_t, k_t, u_t)\rangle dt\Big] \\
&\qquad + \mathbb{E}\Big[\int_0^T \big\langle H_y(t, y_t, Y_t, z_t, Z_t, k_t, u_t, p_t, P_t, q_t, Q_t, V_t), y_t^{v.} - y_t\big\rangle dt\Big] \\
&\qquad + \mathbb{E}\Big[\int_0^T \big\langle H_z(t, y_t, Y_t, z_t, Z_t, k_t, u_t, p_t, P_t, q_t, Q_t, V_t), z_t^{v.} - z_t\big\rangle dt\Big] \\
&\qquad + \mathbb{E}\Big[\int_0^T \big\langle Q_t, \sigma(t, y_t^{v.}, Y_t^{v.}, z_t^{v.}, Z_t^{v.}, k_t^{v.}, v_t) - \sigma(t, y_t, Y_t, z_t, Z_t, k_t, u_t)\big\rangle dt\Big] \\
&\qquad + \mathbb{E}\Big[\int_0^T \int_\Theta \langle V_t(\rho), \varphi(t, y_t^{v.}, Y_t^{v.}, z_t^{v.}, Z_t^{v.}, k_t^{v.}, v_t, \rho) \\
&\qquad - \varphi(t, y_t, Y_t, z_t, Z_t, k_t, u_t, \rho)\rangle \nu(d\rho) dt\Big].
\end{aligned} \tag{15}
$$

Proof Applying integration by parts (Lemma 1) to $\big\langle p_t, Y_t^{v.} - Y_t\big\rangle$ gives

$$
\begin{aligned}
\big\langle p_T, Y_T^{v.} - Y_T\big\rangle &= \big\langle p_0, Y_0^{v.} - Y_0\big\rangle \\
&\quad - \int_0^T \big\langle p_t, f(t, y_t^{v.}, Y_t^{v.}, z_t^{v.}, Z_t^{v.}, k_t^{v.}, v_t) - f(t, y_t, Y_t, z_t, Z_t, k_t, u_t)\big\rangle dt \\
&\quad - \int_0^T \langle p_t, \big(g(t, y_t^{v.}, Y_t^{v.}, z_t^{v.}, Z_t^{v.}, k_t^{v.}, v_t) - g(t, y_t, Y_t, z_t, Z_t, k_t, u_t)\big) \overleftarrow{dB}_t\rangle
\end{aligned}
$$

$$
\begin{aligned}
&+\int_0^T\int_\Theta \left\langle p_t, (k_t^{v.}(\rho)-k_t(\rho))\right\rangle \tilde{N}(d\rho,dt)\\
&+\int_0^T \left\langle p_t, (Z_t^{v.}-Z_t)dW_t\right\rangle - \int_0^T \left\langle Y_t^{v.}-Y_t, q_t\overleftarrow{dB}_t\right\rangle\\
&+\int_0^T \left\langle H_Y(t,y_t,Y_t,z_t,Z_t,k_t,u_t,p_t,P_t,q_t,Q_t,V_t), Y_t^{v.}-Y_t\right\rangle dt\\
&+\int_0^T \left\langle Y_t^{v.}-Y_t, H_Z(t,y_t,Y_t,z_t,Z_t,k_t,u_t,p_t,P_t,q_t,Q_t,V_t)dW_t\right\rangle\\
&+\int_0^T\int_\Theta \langle Y_t^{v.}-Y_t, H_k(t,y_t,Y_t,z_t,Z_t,k_t,u_t,p_t,P_t,q_t,Q_t,V_t)\rangle \tilde{N}(d\rho,dt)\\
&-\int_0^T \left\langle q_t, g(t,y_t^{v.},Y_t^{v.},z_t^{v.},Z_t^{v.},k_t^{v.},v_t)-g(t,y_t,Y_t,z_t,Z_t,k_t,u_t)\right\rangle dt\\
&+\int_0^T \left\langle H_Z(t,y_t,Y_t,z_t,Z_t,k_t,u_t,p_t,P_t,q_t,Q_t,V_t), Z_t^{v.}-Z_t\right\rangle dt\\
&+\int_0^T\int_\Theta \langle H_k(t,y_t,Y_t,z_t,Z_t,k_t,u_t,p_t,P_t,q_t,Q_t,V_t), k_t^{v.}(\rho)-k_t(\rho)\rangle \nu(d\rho)dt.
\end{aligned}
$$

Now by taking the expectation to the above equality, we obtain (14). Similarly

$$
\begin{aligned}
\left\langle P_T, y_T^{v.}-y_T\right\rangle =& \int_0^T \langle P_t, b(t,y_t^{v.},Y_t^{v.},z_t^{v.},Z_t^{v.},k_t^{v.},v_t)-b(t,y_t,Y_t,z_t,Z_t,k_t,u_t)\rangle dt\\
&+\int_0^T \langle P_t, (\sigma(t,y_t^{v.},Y_t^{v.},z_t^{v.},Z_t^{v.},k_t^{v.},v_t)-\sigma(t,y_t,Y_t,z_t,Z_t,k_t,u_t))dW_t\rangle\\
&+\int_0^T\int_\Theta \left\langle y_t^{v.}-y_t, V_t(\rho)\right\rangle \tilde{N}(d\rho,dt) - \int_0^T \left\langle P_t, (z_t^{v.}-z_t)\overleftarrow{dB}_t\right\rangle + \int_0^T \left\langle y_t^{v.}-y_t, Q_t dW_t\right\rangle\\
&+\int_0^T\int_\Theta \langle P_t, (\varphi(t,y_t^{v.},Y_t^{v.},z_t^{v.},Z_t^{v.},k_t^{v.},v_t,\rho)-\varphi(t,y_t,Y_t,z_t,Z_t,k_t,u_t,\rho))\rangle \tilde{N}(d\rho,dt)\\
&+\int_0^T \left\langle y_t^{v.}-y_t, H_z(t,y_t,Y_t,z_t,Z_t,k_t,u_t,p_t,P_t,q_t,Q_t,V_t)\overleftarrow{dB}_t\right\rangle\\
&+\int_0^T \left\langle H_y(t,y_t,Y_t,z_t,Z_t,k_t,u_t,p_t,P_t,q_t,Q_t,V_t), y_t^{v.}-y_t\right\rangle dt\\
&+\int_0^T \left\langle H_z(t,y_t,Y_t,z_t,Z_t,k_t,u_t,p_t,P_t,q_t,Q_t,V_t), z_t^{v.}-z_t\right\rangle dt\\
&+\int_0^T \langle Q_t, \sigma(t,y_t^{v.},Y_t^{v.},z_t^{v.},Z_t^{v.},k_t^{v.},v_t)-\sigma(t,y_t,Y_t,z_t,Z_t,k_t,u_t)\rangle dt\\
&+\int_0^T\int_\Theta \langle V_t(\rho), \varphi(t,y_t^{v.},Y_t^{v.},z_t^{v.},Z_t^{v.},k_t^{v.},v_t,\rho)-\varphi(t,y_t,Y_t,z_t,Z_t,k_t,u_t,\rho)\rangle \nu(d\rho)dt.
\end{aligned}
$$

By taking the expectation to this equality (15) holds.

The remaining is devoted to completing the proof of Theorem 3.

Proof (*Proof of Theorem* 3) Observe first from (6) that

$$\begin{aligned}
&\ell(t, y_t^{v.}, Y_t^{v.}, z_t^{v.}, Z_t^{v.}, k_t^{v.}, v_t) - \ell(t, y_t, Y_t, z_t, Z_t, k_t, u_t) \\
&\quad = -\big(H(t, y_t^{v.}, Y_t^{v.}, z_t^{v.}, Z_t^{v.}, k_t^{v.}, v_t, p_t, q_t, P_t, Q_t, V_t) \\
&\quad - H(t, y_t, Y_t, z_t, Z_t, k_t, u_t, p_t, q_t, P_t, Q_t, V_t)\big) \\
&\quad + \big\langle p_t, f(t, y_t^{v.}, Y_t^{v.}, z_t^{v.}, Z_t^{v.}, k_t^{v.}, v_t) - f(t, y_t, Y_t, z_t, Z_t, k_t, u_t)\big\rangle \\
&\quad - \langle P_t, b(t, y_t^{v.}, Y_t^{v.}, z_t^{v.}, Z_t^{v.}, k_t^{v.}, v_t) - b(t, y_t, Y_t, z_t, Z_t, k_t, u_t)\rangle \\
&\quad + \langle q_t, g(t, y_t^{v.}, Y_t^{v.}, z_t^{v.}, Z_t^{v.}, k_t^{v.}, v_t) - g(t, y_t, Y_t, z_t, Z_t, k_t, u_t)\rangle \\
&\quad - \big\langle Q_t, \sigma(t, y_t^{v.}, Y_t^{v.}, z_t^{v.}, Z_t^{v.}, k_t^{v.}, v_t) - \sigma(t, y_t, Y_t, z_t, Z_t, k_t, u_t)\big\rangle \\
&\quad - \int_\Theta \langle V_t(\rho), \varphi(t, y_t^{v.}, Y_t^{v.}, z_t^{v.}, Z_t^{v.}, k_t^{v.}, v_t, \rho) \\
&\quad - \varphi(t, y_t, Y_t, z_t, Z_t, k_t, u_t, \rho)\rangle \nu(d\rho).
\end{aligned} \tag{16}$$

Next apply Lemma 3 and (16) in Lemma 2 to find that

$$\begin{aligned}
&J(v.) - J(u.) \\
&\geq \mathbb{E}\Big[\int_0^T \big\langle H_y(t, y_t, Y_t, z_t, Z_t, k_t, u_t, p_t, P_t, q_t, Q_t, V_t), y_t^{v.} - y_t\big\rangle dt\Big] \\
&+ \mathbb{E}\Big[\int_0^T \big\langle H_Y(t, y_t, Y_t, z_t, Z_t, k_t, u_t, p_t, P_t, q_t, Q_t, V_t), Y_t^{v.} - Y_t\big\rangle dt\Big] \\
&+ \mathbb{E}\Big[\int_0^T \big\langle H_z(t, y_t, Y_t, z_t, Z_t, k_t, u_t, p_t, P_t, q_t, Q_t, V_t), z_t^{v.} - z_t\big\rangle dt\Big] \\
&+ \mathbb{E}\Big[\int_0^T \big\langle H_Z(t, y_t, Y_t, z_t, Z_t, k_t, u_t, p_t, P_t, q_t, Q_t, V_t), Z_t^{v.} - Z_t\big\rangle dt\Big] \\
&+ \mathbb{E}\Big[\int_0^T \int_\Theta \langle H_k(t, y_t, Y_t, z_t, Z_t, k_t, u_t, p_t, P_t, q_t, Q_t, V_t), k_t^{v.}(\rho) - k_t(\rho)\rangle \nu(d\rho) dt\Big] \\
&- \mathbb{E}\Big[\int_0^T \big(H(t, y_t^{v.}, Y_t^{v.}, z_t^{v.}, Z_t^{v.}, k_t^{v.}, v_t, p_t, P_t, q_t, Q_t, V_t) \\
&- H(t, y_t, Y_t, z_t, Z_t, k_t, u_t, p_t, P_t, q_t, Q_t, V_t)\big) dt\Big].
\end{aligned} \tag{17}$$

Here we have used the formula $h(\omega, x) := c\,Rx + \xi(\omega), x \in \mathbb{R}^n$, to get the cancelation

$$\mathbb{E}\big[\langle c\,R^* p_T, y_T^{v.} - y_T\rangle\big] - \mathbb{E}\big[\langle p_T, Y_T^{v.} - Y_T\rangle\big] = 0$$

resulting from (13) of Lemma 2 and (15) of Lemma 3.

On the other hand, from the concavity condition (ii) of the mapping

$$(y, Y, z, Z, k, v) \mapsto H(t, y, Y, z, Z, k, v, p_t, P_t, q_t, Q_t, V_t)$$

it follows that

$$\begin{aligned}
&H(t, y_t^{v.}, Y_t^{v.}, z_t^{v.}, Z_t^{v.}, k_t^{v.}, v_t, p_t, P_t, q_t, Q_t, V_t) - H(t, y_t, Y_t, z_t, Z_t, k_t, u_t, p_t, P_t, q_t, Q_t, V_t) \\
&\quad \leq \big\langle H_y(t, y_t, Y_t, z_t, Z_t, k_t, u_t, p_t, P_t, q_t, Q_t, V_t), y_t^{v.} - y_t\big\rangle \\
&\quad + \big\langle H_Y(t, y_t, Y_t, z_t, Z_t, k_t, u_t, p_t, P_t, q_t, Q_t, V_t), Y_t^{v.} - Y_t\big\rangle
\end{aligned}$$

$$
\begin{aligned}
&+ \left\langle H_z(t, y_t, Y_t, z_t, Z_t, k_t, u_t, p_t, P_t, q_t, Q_t, V_t), z_t^{v.} - z_t \right\rangle \\
&+ \left\langle H_Z(t, y_t, Y_t, z_t, Z_t, k_t, u_t, p_t, P_t, q_t, Q_t, V_t), Z_t^{v.} - Z_t \right\rangle \\
&+ \int_\Theta \left\langle H_k(t, y_t, Y_t, z_t, Z_t, k_t, u_t, p_t, P_t, q_t, Q_t, V_t), k_t^{v.}(\rho) - k_t(\rho) \right\rangle \nu(d\rho) \\
&+ \left\langle H_v(t, y_t, Y_t, z_t, Z_t, k_t, u_t, p_t, P_t, q_t, Q_t, V_t), v_t - u_t \right\rangle .
\end{aligned}
$$

In particular,

$$
\begin{aligned}
&- \langle H_v(t, y_t, Y_t, z_t, Z_t, k_t, u_t, p_t, P_t, q_t, Q_t, V_t), v_t - u_t \rangle \\
&\quad \leq \left\langle H_y(t, y_t, Y_t, z_t, Z_t, k_t, u_t, p_t, P_t, q_t, Q_t, V_t), y_t^{v.} - y_t \right\rangle \\
&+ \left\langle H_Y(t, y_t, Y_t, z_t, Z_t, k_t, u_t, p_t, P_t, q_t, Q_t, V_t), Y_t^{v.} - Y_t \right\rangle \\
&+ \left\langle H_z(t, y_t, Y_t, z_t, Z_t, k_t, u_t, p_t, P_t, q_t, Q_t, V_t), z_t^{v.} - z_t \right\rangle \\
&+ \left\langle H_Z(t, y_t, Y_t, z_t, Z_t, k_t, u_t, p_t, P_t, q_t, Q_t, V_t), Z_t^{v.} - Z_t \right\rangle \\
&+ \int_\Theta \left\langle H_k(t, y_t, Y_t, z_t, Z_t, k_t, u_t, p_t, P_t, q_t, Q_t, V_t), k_t^{v.}(\rho) - k_t(\rho) \right\rangle \nu(d\rho) \\
&- [H(t, y_t^{v.}, Y_t^{v.}, z_t^{v.}, Z_t^{v.}, k_t^{v.}, v_t, p_t, P_t, q_t, Q_t, V_t) \\
&\qquad\qquad - H(t, y_t, Y_t, z_t, Z_t, k_t, u_t, p_t, P_t, q_t, Q_t, V_t)].
\end{aligned}
$$

Now by applying this latter result in (17) we obtain

$$
J(v.) - J(u.) \geq -\mathbb{E}\Big[\int_0^T \langle H_v(t, y_t, Y_t, z_t, Z_t, k_t, u_t, p_t, P_t, q_t, Q_t, V_t), v_t - u_t \rangle \, dt\Big]. \tag{18}
$$

On the other hand, the maximum condition (iii) yields

$$
\langle H_v(t, y_t, Y_t, z_t, Z_t, k_t, u_t, p_t, P_t, q_t, Q_t, V_t), v_t - u_t \rangle \leq 0.
$$

Hence (18) becomes

$$
J(v.) - J(u.) \geq 0.
$$

Since $v.$ is an arbitrary element of $\mathscr{U}_{ad}$, this inequality completes the proof if we recall (5).

Acknowledgments This work is supported by the Science College Research Center at Qassim University, project no. SR-D-012-1958. It is also supported by the Algerian CNEPRU project no. B01420130137.

References

1. Ahmed, N.U., Charalambous, C.D.: Stochastic minimum principle for partially observed systems subject to continuous and jump diffusion processes and driven by relaxed controls, arXiv:1302.3455v1 [math.OC], (2013)
2. Antonelli, F.: Backward-forward stochastic differential equations. Ann. Appl. Probab. **3**, 777–793 (1993)
3. Bahlali, K., Khelfallah, N., Mezerdi, B.: Necessary and sufficient conditions for near-optimality in stochastic control of FBSDEs. Syst. Control Lett. **58**(12), 857–864 (2009)
4. Meng, Q. Sun, Y.: Optimal control problem of fully coupled forward-backward stochastic systems with Poisson jumps under partial information. In: Proceedings at 30th Chinese Control Conference, Yantai, China, 22–24 July 2011, pp. 1317–1322
5. Nualart, D., Pardoux, E.: Stochastic calculus associated with Skokrohod's integral, Lecture Notes in Control and Inf. Sciences **96**, 363–372 (1987)
6. Øksendal, B., Sulem, A.: Maximum principles for optimal control of forward-backward stochastic differential equations with jumps. SIAM J. Control Optim. **48**(5), 2945–2976 (2009)
7. Pardoux, E., Peng, S.: Backward doubly stochastic differential equations and system of quasilinear SPDEs. Prob. Th. & Rel. Fields **98**(2), 209–227 (1994)
8. Peng, S., Shi, Y.: A type of time-symmetric forward-backward stochastic differential equations, C. R. Acad. Sci. Paris Ser. I **336**(1), 773–778 (2003)
9. Peng, S., Wu, Z.: Fully coupled forward-backward stochastic differential equations and applications to optimal control. SIAM J. Control Optim **37**, 825–843 (1999)
10. Shi, J.: Necessary conditions for optimal control of forward-backward stochastic systems with random jumps, International Journal of Stochastic Analysis, **2012**, Article ID 258674, (2012) doi:10.1155/2012/258674
11. Shi, J., Wu, Z.: The maximum principle for fully coupled forward-backward stochastic control system. Acta Automatica Sinica **32**(2), 161–169 (2006)
12. Shi, J., Wu, Z.: Maximum principle for partially-observed optimal control of fully-coupled forward-backward stochastic systems. J. Optim. Theory. Appl. **145**, 543–578 (2010)
13. Shi, J., Wu, Z.: Maximum principle for forward-backward stochastic control system with random jumps and applications to finance. J. Syst. Sci. Complex. **23**(2), 219–231 (2010)
14. Situ, R.: On solution of backward stochastic differential equations with jumps and applications. Stoch. Process. Appl., **66**, 209–236 (1997)
15. Sun, X., Lu, Y.: The property for solutions of the multi-dimensional backward doubly stochastic differential equations with jumps. Chin. J. Appl. Probab. Stat. **24**, 73–82 (2008)
16. Wu, Z.: Maximum principle for optimal control problem of fully coupled forward-backward stochastic systems. Syst. Sci. Math. Sci. **11**(3), 249–259 (1998)
17. Wu, Z.: Forward-backward stochastic differential equations with Brownian motion and Poisson process. Acta Math. Appl. Sinica **15**, 433–443 (1999)
18. Wu, Z., Wang, X.: FBSDE with Poisson process and its application to linear quadratic stochastic optimal control problem with random jumps. Acta Automatica Sinica **29**, 821–826 (2003)
19. Xu, S., Jiang, J.: Maximum principle for forward-backward doubly stochastic differential equations with jumps, [OL], http://www.paper.edu.cn/en_releasepaper/content/4486654
20. Xu, W.: Stochastic maximum principle for optimal control problem of forward and backward system. J. Aust. Math. Soc. B **37**, 172–185 (1995)
21. Yin, J., Situ, R.: On solutions of forward-backward stochastic differential equations with Poisson jumps. Stoch. Anal. Appl. **21**, 1419–1448 (2003)
22. Yong, J.: Finding adapted solutions of forward-backward stochastic differential equations-method of continuation. Prob. Th. & Rel. Fields **107**, 537–572 (1997)
23. Zhang, L., Shi, Y.: General doubly stochastic maximum principle and its applications to optimal control of stochastic partial differential equations, arXiv:1009.6061v3 [math.OC], (2012)

24. Zhu, Q., Shi, Y.: Forward-backward doubly stochastic differential equations with random jumps and stochastic partial differential-integral equations, [OL].[201001-1044], http://www.paper.edu.cn/index.php/default/en-releasepaper/downPaper/201001-1044
25. Zhu, Q., Shi, Y.: Forward-backward doubly stochastic differential equations and related stochastic partial differential equations. Sci. Chin. Math. **55**(12), 2517–2534 (2012)

On the Pathwise Uniqueness of Solutions of One-Dimensional Stochastic Differential Equations with Jumps

Mohsine Benabdallah, Siham Bouhadou and Youssef Ouknine

Abstract We consider one-dimensional stochastic differential equations with jumps in the general case. We introduce new technics based on local time and we prove new results on pathwise uniqueness and comparison theorems. Our approach are very easy to handled. Similar equations without jumps were studied in the same context by [10, 15] and others authors.

Keywords Semimartingale · Local time · Tanaka formula · Pathwise uniqueness

Mathematical Subject Classification 2010 60H10 · 60H20 · 60H99

1 Introduction

Stochastic differential equations play a central role in the theory of stochastic processes and are often used in the modeling of various random processes in nature. Often defined as strong solutions of stochastic differential equations, diffusion processes are widely used in stochastic modeling e.g., Ornstein and Uhlenbeck [11] used their process for the analysis of velocity of a particle in a fluid under the bombardment by molecules. Samuelson [19] introduced geometric Brownian motion for modeling the behavior of financial markets. Also diffusions processes appear e.g., in stochastic population modeling.

M. Benabdallah
Department of Mathematics, Ibn Tofail University, B.P. 133, Kenitra, Morocco
e-mail: bmohsine@gmail.com

S. Bouhadou · Y. Ouknine (✉)
LIBMA Laboratory, Department of Mathematics, Faculty of Sciences Semlalia,
Cadi Ayyad University, P.B.O. 2390, Marrakesh, Morocco
e-mail: ouknine@uca.ma

S. Bouhadou
e-mail: sihambouhadou@gmail.com

M. Eddahbi et al. (eds.), *Statistical Methods and Applications in Insurance and Finance*, Springer Proceedings in Mathematics & Statistics 158,
DOI 10.1007/978-3-319-30417-5_8

In the recent years, jump processes were also used in many fields as flexible models to describe various phenomena. In particular, they are frequently used in financial modeling, it would be nice for the readers to refer to Cont and Tankov [5] and their reference for details about applications. Barndorff-Nielsen [1] proposed the idea for generalizing diffusion processes by means of changing the driving Wiener process by a Lévy process and defined the so-called background driven Ornstein-Uhlenbeck type process.
In the present paper, we consider stochastic differential equations driven by both a Wiener process and a Poisson random measure, and study the question of pathwise uniqueness of this class called stochastic differential equation with jumps:

$$dX_t = \sigma(X_t)dW_t + b(X_t)dt + \int_{\mathbb{R}-0} F(X_{t-}, z)(\mu - \nu)(dz, dt), \quad X_0 = x_0. \quad (1)$$

In fact, the results on pathwise uniqueness of (1) have been obtained under Lipschitz conditions, see Skorohod [20], Ikeda and Watanbe [9], Protter [17]. In absence of jumps, this SDE

$$dX_t = \sigma(X_t)dW_t + b(X_t)dt,$$

with non-Lipschitz coefficient were considered by several authors. There were many works which discuss under which conditions on b and σ, we have the existence of strong solutions of stochastic differential equations. In the case when the equation is one-dimensional and σ is not degenerated, several results have been obtained by Ouknine [12–14]. For SDEs which involve local times of unknown process, the most general result is given by Rutkowski [18] where he showed the so called (LT) condition is sufficient to have pathwise uniqueness. So, the purpose of this paper is to give the analogue of this condition for the one dimensional SDE with jumps which concerns the couple of coefficients σ and F.

This paper is arranged as follows. In Sect. 2, we recall the definition of (LT) condition introduced by Barlow and Perkins [2], and we introduce the new definition of local time condition ($\mathscr{L}\mathscr{T}$). In Sect. 3, we give some sufficient assumptions which ensure this condition.

2 Preliminaries

On some stochastic basis $(\Omega, \mathscr{A}, \mathbb{F} = (\mathscr{F}_t)_{t\geq 0}, P)$, we consider one-dimensional $\mathbb{F}$-Brownian motion $W = (W_t)_{t\geq 0}$ and a $\mathbb{F}$-Poisson point process $\{p_t\}$. Let $\mu(ds, dy)$ be the Poisson random measure associated with $\{p_t\}$. We suppose that μ and W are independent. The random measure $\mu(ds, dy)$ has deterministic intensity

$$\nu(ds, dy) = ds\lambda(dy) \quad \text{on} \quad (0, \infty) \times \mathbb{R} - \{0\}$$

where λ is σ-finite measure on $\mathbb{R} - \{0\}$, satisfying

$$\int_{\mathbb{R}-0} (y \wedge 1)^2 \lambda(dy) < \infty.$$

A solution of (1) is any process $X = (X_t)_{t \geq 0}$ on $(\Omega, \mathscr{A}, \mathbb{F}, P)$ satisfying (i) and (ii) below. We say that the pathwise uniqueness of solutions for (1) holds if whenever X and X' are any two solutions defined on the same stochastic basis $(\Omega, \mathscr{A}, \mathbb{F}, P)$ with the same $\mathbb{F}$-Brownian motion $W = (W_t)_{t \geq 0}$ and the same $\mathbb{F}$-Poisson point process $\mu(ds, dy)$ such that $X_0 = X'_0$ a.s., then $X_t = X'_t$ for all $t \geq 0$ a.s.

We present, following Protter [17] the notion of local time of a semimartingale and some of its properties. If X is a general càdlàg semimartingale, let ΔX denote the process $\Delta X_t = X_t - X_{t-}$.

We recall the quadratic variation process of X is defined by

$$[X]_t = X_t^2 - 2 \int_0^t X_{s-} dX_s.$$

The local time at a of X, denoted $L_t^a = L_t^a(X)$ is defined to be the process given by

$$\begin{aligned} L_t^a &= |X_t - a| - |X_0 - a| - \int_0^t \operatorname{sign}(X_{s-} - a) dX_s \\ &\quad - \sum_{0<s\leq t} \{|X_s - a| - |X_{s-} - a| - \operatorname{sign}(X_{s-} - a)\Delta X_s\}. \end{aligned}$$

The local time gives a generalization of Itô's formula: if f is the difference of two convex functions and f' is its left derivative and let μ be the signed measure which is the second derivative of f. Thus we have

$$\begin{aligned} f(X_t) &= f(X_0) + \int_{0^+}^t f'(X_{s-}) dX_s + \sum_{0<s\leq t} \{f(X_s) - f(X_{s-}) - f'(X_{s-})\Delta X_s\} \\ &\quad + \frac{1}{2} \int_{-\infty}^{\infty} L_t^a \mu(da). \end{aligned}$$

We introduce the local time slanted $\mathscr{L}_t^a(X)$ of a semimartingale X by

$$\mathscr{L}_t^a = |X_t - a| - |X_0 - a| - \int_0^t \operatorname{sign}(X_{s-} - a) dX_s.$$

We remark that if f is the difference of two convex functions, we have

$$f(X_t) = f(X_0) + \int_{0^+}^t f'(X_{s-}) dX_s + \frac{1}{2} \int_{-\infty}^{\infty} \mathscr{L}_t^a \mu(da).$$

Finally, we indicate the formula of occupation density, if f is a bounded Borel measurable function, then, *a.s.*

$$\int_{-\infty}^{\infty} L_t^a f(a)da = \int_0^t f(X_{s-})d\langle X\rangle_s,$$

where $\langle X\rangle$ is the continuous part of $[X]$.
We apply this notation:
For all x and y in $\mathbb{R}$, $x \wedge y = \inf(x, y)$ and $x \vee y = \sup(x, y)$.
Now, we introduce two definitions of the (LT) condition, the first concerns the coefficient σ uniquely and the second concerns the couple of coefficients (σ, F) that will help us to get the pathwise uniqueness of Eq. (1).

Definition 1 We say that a coefficient σ of Eq. (1) satisfies the (LT) condition if for two solutions X^1 and X^2 of (1), then

$$\forall t \geq 0 \;\; L_t^0(X^1 - X^2) = 0. \tag{2}$$

Now, we define the $(\mathscr{L}\mathscr{T})$ condition concerning the couple of coefficient (σ, F) of Eq. (1).

Definition 2 We say that the coefficients (σ, F) of Eq. (1) satisfy $(\mathscr{L}\mathscr{T})$ condition if for two solutions X^1 and X^2 of (1), then

$$\forall t \geq 0 \;\; \mathscr{L}_t^0(X^1 - X^2) = 0. \tag{3}$$

We can remark that if the coefficients of (1) verify the $(\mathscr{L}\mathscr{T})$ condition then they verify the (LT) condition too which was used by several authors (Le Gall [10], Ouknine [15]) and which permits to prove the pathwise uniqueness of solutions of one-dimensional stochastic differential equations without jumps.

3 The Pathwise Uniqueness Property for Eq. (1)

3.1 Main Result

Throughout this paragraph, we make the following assumption on the coefficients in Eq. (1):

(A) The functions σ and b are measurable and bounded.
(B) $|b(x) - b(y)| \leq c\,|x - y|$ for all x, y.
(C) $|b(x)|^2 + |\sigma(x)|^2 + \int |F(x, z)|^2\, \lambda(dz) + \int |F(x, z)|\, \lambda(dz) \leq c(1 + |x|^2)$ for all x.

Theorem 1 *If* (σ, F) *verify* $(\mathscr{L}\mathscr{T})$ *condition, then the solution to* (1) *is pathwise unique.*

Proof We need the following Lemma to prove that if pathwise uniqueness holds for (1) before the first big jump(for example $|\Delta X| \geq 1$), then pathwise uniqueness holds for every $t \geq 0$. This allows to consider Eq. (1) with only the small jumps.

Lemma 1 *We have pathwise uniqueness for Eq.* (1) *with only the small jumps, then pathwise uniqueness holds for Eq.* (1) *for general case.*

Proof Let X and Y two solutions of Eq. (1) with the same initial value. Let S_1 be the first time when the big jump happens ($|\Delta X| \geq 1$), then we have

$$X_t = Y_t \quad \text{a.s. for } t \in [0, S_1)\,.$$

We have also $X_{S_1} = X_{S_1^-} + F(X_{S_1^-}, \Delta X_{S_1}) = Y_{S_1^-} + F(Y_{S_1^-}, \Delta Y_{S_1}) = Y_{S_1}$. We consider the filtration $\mathbb{F}^{S_1} = (\mathscr{F}_{S_1+t})_{t\geq 0}$, the $\mathbb{F}^{S_1}$-Brownian motion $W^{S_1} = (W_{S_1+t} - W_{S_1})_{t\geq 0}$ and a $\mathbb{F}^{S_1}$-Poisson point process $\mu^{S_1}(ds, dy)$ with intensity $\nu^{S_1} = ds\lambda(dy)$ on in $(0, \infty) \times \mathbb{R}$ defined by $\mu^{S_1}([0, t] \times \cdot) = \mu([S_1, S_1 + t] \times \cdot)$. We consider the new equation:

$$dX_t^{S_1} = \sigma(X_t^{S_1})dW_t^{S_1} + b(X_t^{S_1})dt + \int F(X_{t-}^{S_1}, z)(\mu^{S_1} - \nu^{S_1})(dz, dt), \quad t \geq 0. \quad (4)$$

We consider the processes X^{S_1} and Y^{S_1} defined by $X_t^{S_1} = X_{S_1+t},\ t \geq 0$ and $Y_t^{S_1} = Y_{S_1+t},\ t \geq 0$. Then X^{S_1} and Y^{S_1} are solutions of (4) with the same initial condition and by the hypothesis of pathwise uniqueness property of stochastic differential equation without the big jumps, they are equal until the first big jump when happens at time S_2. We have also

$$X_t^{S_1} = Y_t^{S_1} \quad \text{a.s. for } t \in [0, S_2 - S_1)\,.$$

This implies

$$X_t = Y_t \quad \text{a.s. for } t \in [0, S_2)\,.$$

We repeat the same process, we obtain the pathwise uniqueness for Eq. (1) for the general case.

To complete the proof, we use the Tanaka's formula. Let X^1 and X^2 be two solutions to (1), then

$$|X_t^1 - X_t^2| = \int_0^t \operatorname{sign}(X_{s-}^1 - X_{s-}^2)d(X_s^1 - X_s^2). \quad (5)$$

Now, we have

$$\begin{aligned} d(X_t^1 - X_t^2) &= (\sigma(X_t^1) - \sigma(X_t^2))dW_t + (b(X_t^1) - b(X_t^2))dt \\ &\quad + \int (F(X_{t-}^1, z) - F(X_{t-}^2, z))(\mu - \nu)(dt, dz). \end{aligned} \quad (6)$$

If we substitute (6) in (5), we obtain

$$\begin{aligned}|X_t^1 - X_t^2| &= \int_0^t \operatorname{sign}(X_{s-}^1 - X_{s-}^2)(\sigma(X_s^1) - \sigma(X_s^2))dW_s \\ &+ \int_0^t \operatorname{sign}(X_{s-}^1 - X_{s-}^2)(b(X_s^1) - b(X_s^2))ds \\ &+ \int_0^t \operatorname{sign}(X_{s-}^1 - X_{s-}^2) \int (F(X_{s-}^1, z) - F(X_{s-}^2, z))(\mu - \nu)(ds, dz).\end{aligned}$$

The first and the third terms on the right hand sides are martingales and all terms on the right hand side are integrable.
We can prove, if we note $I_2(t)$ the second term on the right hand side and using the fact that b is Lipschitz, that

$$|I_2(t)| \le c \int_0^t |X_s^1 - X_s^2| \, ds. \tag{7}$$

Finally, we obtain

$$\mathbb{E}\left(|X_t^1 - X_t^2|\right) \le c \left(\int_0^t \mathbb{E}\,|X_s^1 - X_s^2| \, ds\right)$$

and Gronwall's lemma implies that $X^1 \equiv X^2$.
In the following, we present a sufficient assumption to get the $(\mathcal{LT})$ condition.

Proposition 1 *Suppose that there exist a sequence of non-negative and twice continuously differentiable functions* $\{\phi_n\}$ *with the following properties:*

(a) $\phi_n(z) \uparrow |z|$ *as* $n \to \infty$.
(b) $|\phi_n'(z)| \le 1$ *for all* z.
(c) $\phi_n''(z) \ge 0$ *for* $z \in \mathbb{R}$ *and as* $n \to \infty$,

$$\phi_n''(x - y)\,[\sigma(x) - \sigma(y)]^2 \to 0 \text{ uniformly in } |x|, |y| \le m.$$

(d) as $n \to \infty$

$$\begin{aligned}\int &[\phi_n(x + F(x, z) - y - F(y, z)) - \phi_n(x - y) - \phi_n'(x - y)(F(x, z) \\ &- F(y, z))]\,\lambda(dz)\end{aligned}$$

converges to 0 *uniformly in* $|x|, |y| \le m$;

then (σ, F) *verify* $(\mathcal{LT})$ *condition.*

Proof Let X^1 and X^2 be two solutions of (1) defined on the same stochastic basis $(\Omega, \mathcal{A}, \mathbb{F}, P)$ with the same $\mathbb{F}$-Brownian motion $W = (W_t)_{t \ge 0}$ and the same $\mathbb{F}$-Poisson point process $\mu(ds, dy)$ such that $X_0^1 = X_0^2$ a.s., we set:

$$I_n(s) = \int \{\phi_n((X^1_{s-} - X^2_{s-}) + [F(X^1_{s-}, z) - F(X^2_{s-}, z)])$$

$$- \phi_n(X^1_{s-} - X^2_{s-}) - [F(X^1_{s-}, z) - F(X^2_{s-}, z)]\, \phi_n'(X^1_{s-} - X^2_{s-})\}\, \lambda(dz).$$

Let $\tau_m = \inf\{t \geq 0 : |X_1(t)| \geq m \text{ or } |X_2(t)| \geq m\}$. By application of Itô's formula, we have

$$\begin{aligned}
\int_0^{t\wedge\tau_m} I_n(s)ds &= \phi_n(X^1_{t\wedge\tau_m} - X^2_{t\wedge\tau_m}) - \phi_n(0) \\
&- \int_0^{t\wedge\tau_m} \phi_n'(X^1_{s-} - X^2_{s-})(b(X^1_{s-}) - b(X^2_{s-}))ds \\
&- \int_0^{t\wedge\tau_m} \phi_n'(X^1_{s-} - X^2_{s-})(\sigma(X^1_{s-}) - \sigma(X^2_{s-}))dW_s \\
&- \frac{1}{2}\int_0^{t\wedge\tau_m} \phi_n''(X^1_{s-} - X^2_{s-})(\sigma(X^1_{s-}) - \sigma(X^2_{s-}))^2 ds \\
&- \int_0^{t\wedge\tau_m} \int [\phi_n((X^1_{s-} - X^2_{s-}) + (F(X^1_{s-}, z) - F(X^2_{s-}, z)) \\
&- \phi_n(X^1_{s-} - X^2_{s-})]\,(\mu - \nu)(dz, ds).
\end{aligned}$$

According to the assumptions (a), (c), (d), the term on the left hand side, the second and the fifth terms on the right hand side tend to zero. We obtain:

$$\begin{aligned}
0 &= \left|(X^1_{t\wedge\tau_m} - X^2_{t\wedge\tau_m})\right| \\
&- \textstyle\int_0^{t\wedge\tau_m} \operatorname{sign}(X^1_{s-} - X^2_{s-})(b(X^1_{s-}) - b(X^2_{s-}))ds \\
&- \textstyle\int_0^{t\wedge\tau_m} \operatorname{sign}(X^1_{s-} - X^2_{s-})(\sigma(X^1_{s-}) - \sigma(X^2_{s-}))dW_s \\
&- \textstyle\int_0^{t\wedge\tau_m} \int \left[\left|X^1_{s-} - X^2_{s-} + (F(X^1_{s-}, z) - F(X^2_{s-}, z))\right|\right. \\
&- \left.\left|X^1_{s-} - X^2_{s-}\right|\right](\mu - \nu)(dz, ds).
\end{aligned}$$

We note that this leads to the formula of Tanaka which lacks a term that can only be zero in this case. We obtain

$$\left|X^1_{t\wedge\tau_m} - X^2_{t\wedge\tau_m}\right| - \int_0^{t\wedge\tau_m} \operatorname{sign}(X^1_{s-} - X^2_{s-})d(X^1_s - X^2_s) = 0.$$

Since $\tau_m \to \infty$ as $m \to \infty$, we obtain

$$\mathscr{L}^0_t(X^1 - X^2) = |X^1_t - X^2_t| - \int_0^t \operatorname{sign}(X^1_{s-} - X^2_{s-})d(X^1_s - X^2_s) = 0.$$

Thus, (σ, F) verify $(\mathscr{L}\mathscr{T})$ condition.

Now, we give several conditions which ensure $(\mathscr{L}\mathscr{T})$ condition, in particular the coefficients can be discontinuous. First, we need the following Lemma:

Lemma 2 *Suppose that the condition* (C) *is satisfied. Let* X *be a solution of* (1), *then there exists a version of* $(a, t, w) \to L_t^a(X)$ *which is everywhere jointly continuous in* t *and right continuous in* a.

Proof To prove the lemma, we may verify that: $\sum_{s\leq t} |\Delta X_s| < \infty$ a.s.; see e.g. Theorem 75, Protter [17]. So, it is sufficient to prove that $\sum_{s\leq t} |\Delta X_s| < \infty$. Let $\alpha_t = \sum_{s\leq t} |\Delta X_s|$. From (1) and condition (C):

$$\begin{aligned}\mathbb{E}(\alpha_t) &= \mathbb{E}\left(\sum_{s\leq t} |\Delta p(t)|\right) = \mathbb{E}\left(\int_0^t \int |F(X_{s-}, u)|\mu(ds, du)\right)\\ &= \mathbb{E}\left(\int_0^t \int |F(X_{s-}, u)|\lambda(du)ds\right)\\ &< \infty\end{aligned}$$

Corollary 1 *If* σ *and* F *verify the following condition:*

$$|\sigma(x) - \sigma(y)|^2 + \int_{|z|<1} (F(x, z) - F(y, z))^2\lambda(dz) \leq h(|x - y|) \text{ for all } x, y$$

where $h : [0, \infty) \to [0, \infty)$ *is continuous and nondecreasing,* $h(0) = 0$, $h(x) > 0$ *for* $x > 0$, *and*

$$\int_0^\varepsilon \frac{du}{h(u)} = \infty \text{ for every } \varepsilon > 0,$$

and if $x \to x + F(x, z)$ *is nondecreasing in a neighborhood of* 0, $\lambda(dz)$ *a.e. then* (σ, F) *verify* $(\mathscr{L}\mathscr{T})$ *condition.*

Proof The main idea is to use the Tanaka formula but with the function $x \to x^+$. Let X^1 and X^2 be two solutions to (1), then

$$\begin{aligned}(X_t^1 - X_t^2)^+ = &\int_0^t \mathbf{1}_{(X_{s-}^1 - X_{s-}^2 > 0)} d(X_s^1 - X_s^2)\\ &+ \sum_{0<s\leq t} \mathbf{1}_{(X_{s-}^1 - X_{s-}^2 > 0)} \left(X_s^1 - X_s^2\right)^-\\ &+ \sum_{0<s\leq t} \mathbf{1}_{(X_{s-}^1 - X_{s-}^2 \leq 0)} \left(X_s^1 - X_s^2\right)^+\\ &+ \frac{1}{2} L_t^0(X^1 - X^2).\end{aligned}$$

In order that the $(\mathscr{L}\mathscr{T})$ condition be realized, we can prove that the second and the third term on the right hand side is zero and that $L_t^0(X^1 - X^2) = 0$. By the first hypothesis on σ and F and Itô's formula, we obtain:

$$
\begin{aligned}
&\int_0^t 1_{\{X^1_{s-}-X^2_{s-}>0\}} \frac{d\langle X^1 - X^2\rangle_s}{h(X^1_{s-} - X^2_{s-})} \\
&\leq \int_0^t \frac{(\sigma(X^1_s) - \sigma(X^2_s))^2 + \int_{|z|<1}(F(X^1_s, z) - F(X^2_s, z))^2\lambda(dz)}{h(X^1_s - X^2_s)} 1_{\{X^1_s - X^2_s>0\}}\, ds \\
&\leq t.
\end{aligned}
$$

Hence, by the occupation time formula, we get:

$$
\int_0^t 1_{\{X^1_{s-}-X^2_{s-}>0\}} \frac{d\langle X^1 - X^2\rangle_s}{h(X^1_{s-} - X^2_{s-})} = \int_{0+}^{+\infty} \frac{da}{h(a)} L^a_t(X^1 - X^2) < \infty.
$$

Thus, by Lemma 2, $a \to L^a_t(X^1 - X^2)$ is right continuous. This argument combined with this condition $\int_{0+}^{\varepsilon} \frac{du}{h(u)} = \infty$ for some $\varepsilon > 0$ imply that $L^0_t(X^1 - X^2) = 0$. Now, we prove uniquely the nullity of the second term, the other is proved by the same method. We can suppose that $x \to F(x, z) + x$ is nondecreasing in a neighborhood of 0, $\lambda(dz)$ *a.e.*. We have

$$
\begin{aligned}
&\textstyle\sum_{0<s\leq t} \mathbf{1}_{(X^1_{s-}-X^2_{s-}>0)} \left(X^1_s - X^2_s\right)^- \\
&= \textstyle\sum_{0<s\leq t} \mathbf{1}_{(X^1_{s-}-X^2_{s-}>0)} \left(\Delta X^1_s - \Delta X^2_s + X^1_{s-} - X^2_{s-}\right)^- \qquad (8) \\
&= \int_0^t \int \mathbf{1}_{(X^1_{s-}-X^2_{s-}>0)} \left[(F(X^1_{s-}, z) + X^1_{s-}) - (F(X^2_{s-}, z) + X^2_{s-})\right]^- \mu(ds, dz).
\end{aligned}
$$

Since $x \to x + F(x, z)$ is nondecreasing, the right hand size of (8) is 0. We obtain the desired result.

This corollary generalizes the results of Fu-Li [8] F is weaker than [8]. We present a second corollary in the spirit of Nakao (Le Gall [10]) result but for discontinuous SDE.

Corollary 2 *If* $\sigma = 0$ *and* F *verifies the following condition:*

$$
\int_{|z|<1} (F(x, z) - F(y, z))^2\lambda(dz) \leq |f(x) - f(y)| \text{ for all } x, y
$$

where $f : \mathbb{R} \to \mathbb{R}$ *is of bounded variation, if there exists* $\varepsilon > 0$ *such that* $F > \varepsilon$ *and if* $x \to x + F(x, z)$ *is nondecreasing in a neighborhood of* 0, $\lambda(dz)$ *a.e. then* F *verifies* $(\mathscr{L}\mathscr{T})$ *condition.*

Proof Without loss of generality, we shall prove the statement for an increasing function f. Let us first show that $L^0_\cdot(X - Y) \equiv 0$, whenever X and Y denote any two solutions of the SDE (1). By Lemma 2, we get the right continuity of $L^0_\cdot$. So, it is enough to prove that, for any $t \geq 0$,

$$\int_{0^+}^{+\infty} \frac{L_t^a(X-Y)}{a} da < +\infty.$$

Indeed, using the density occupation formula we can write,

$$\begin{aligned}\int_{0^+}^{+\infty} \frac{L_t^a(X-Y)}{a} da &= \int_0^t \frac{d\langle X-Y\rangle_s}{X_{s-}-Y_{s-}} 1_{\{X_{s-}-Y_{s-}>0\}} \\ &= \int_0^t \frac{\int_{|z|<1}(F(X_s,z)-F(Y_s,z))^2\,\lambda(dz)}{X_s-Y_s} 1_{\{X_s-Y_s>0\}} ds.\end{aligned}$$

Applying the assumption of the corollary, we obtain

$$\begin{aligned}&\int_0^t \frac{\int_{|z|<1}(F(X_s,z)-F(Y_s,z))^2\,\lambda(dz)}{X_s-Y_s} 1_{\{X_s-Y_s>0\}} ds \\ &\leq \int_0^t \frac{|f(X_s)-f(Y_s)|}{X_s-Y_s} 1_{\{X_s-Y_s>0\}} ds.\end{aligned}$$

As a consequence,

$$\mathbb{E}\left[\int_{0^+}^{+\infty} \frac{L_t^a(X-Y)}{a} da\right] \leq \mathbb{E}\left[\int_0^t \frac{|f(X_s)-f(Y_s)|}{X_s-Y_s} 1_{\{X_s-Y_s>0\}} ds\right]. \tag{9}$$

Now, by a localization argument $\|f\|_\infty := \sup_x |f(x)| < \infty$.
Let θ_n denote the standard positive regularizing mollifiers sequence, and define

$$f_n(x) = (\widetilde{f}(.) * \theta_n)(x) \qquad \text{for } x \in \mathbb{R}, \quad n \in \mathbb{N}^*,$$

where $\widetilde{f}$ is any real function such that $\widetilde{f}(x) = f(x)$ if $|x| \leq M$ and 0 if $|x| \geq M+1$. Note that f_n are increasing functions, with support contained in $[-M-1, M+1]$ such that

$$\sup_n \sup_x |f_n(x)| \leq \|f\|_\infty \text{ and } f_n(x) \to f(x) \text{ for every } x \in D^c, \ |x| \leq M,$$

where D is the denumerable set of discontinuous points of the function f. If we denote
$Z_t^\alpha = \alpha X_t + (1-\alpha)Y_t$, then, using successively Fatou's Lemma, the intermediate value theorem and $F > \varepsilon$. We get,

$$\begin{aligned}&\mathbb{E}\left[\int_0^t \frac{(f(X_s)-f(Y_s))}{X_s-Y_s} 1_{\{X_s-Y_s>0\}} ds\right] \\ &\leq \liminf_{n\to+\infty} \mathbb{E}\left[\int_0^t \frac{(f_n(X_s)-f_n(Y_s))}{X_s-Y_s} 1_{\{X_s-Y_s>0\}} ds\right]\end{aligned}$$

$$= \liminf_{n\to+\infty} \mathbb{E}\left[\int_0^t \int_0^1 f_n'(\alpha X_s + (1-\alpha)Y_s) d\alpha ds\right]$$
$$= \liminf_{n\to+\infty} \int_0^1 d\alpha \mathbb{E}\left[\int_0^t f_n'(Z_s^\alpha) ds\right]$$
$$\leq \frac{1}{\varepsilon^2} \liminf_{n\to+\infty} \int_0^1 d\alpha \mathbb{E}\left[\int_0^t f_n'(Z_s^\alpha) d\langle Z^\alpha\rangle_s\right].$$

Note that we have used in the first inequality the fact that

$$\int_0^t P[(X_{s-} \in D) \cup (Y_{s-} \in D)] ds = 0.$$

This last equality is a consequence of the occupation time formula. Hence,

$$\mathbb{E}\left[\int_0^{+\infty} \frac{L_t^a(X-Y)}{a} da\right] \leq \frac{1}{\varepsilon^2} \liminf_{n\to+\infty} \mathbb{E}\left[\int_{\mathbb{R}} \int_0^1 f_n'(a) L_t^a(Z^\alpha) d\alpha da\right].$$

Therefore, we obtain

$$\mathbb{E}\left[\int_{0+}^{+\infty} \frac{L_t^a(X-Y)}{a} da\right] \leq C \sup_{\alpha\in[0,1], a\in\mathbb{R}} \mathbb{E}\left[L_t^a(Z^\alpha)\right] \int_{\mathbb{R}} f_n'(a) da$$

where $C > 0$ is a generic constant. Suppose now we can show

$$\sup_{\alpha\in[0,1], a\in\mathbb{R}} \mathbb{E}\left[L_t^a(Z^\alpha)\right] < \infty, \tag{10}$$

this implies that

$$\mathbb{E}\left[\int_{0+}^{+\infty} \frac{L_t^a(X-Y)}{a} da\right] \leq C\|f\|_\infty.$$

Hence $L_\cdot^0(X-Y) \equiv 0$. By Tanaka's formula, we obtain that $|X_\cdot - Y_\cdot|$ is a local martingale, thus also a non-negative supermartingale, with $|X_0 - Y_0| = 0$, and consequently, X and Y are indistinguishable.

The property (10) is checked by standard methods: with the help of the Tanaka's formula

$$|Z_t^\alpha - a| = |Z_0^\alpha - a| + \int_0^t sgn(Z_s^\alpha - a) dZ_s^\alpha + L_t^a(Z^\alpha)$$

and the Itô isometry, we get

$$\mathbb{E}(L_t^a(Z^\alpha)) \leq \mathbb{E}(|Z_t^\alpha - Z_0^\alpha|) + \mathbb{E}(\langle Z^\alpha\rangle_t)^{\frac{1}{2}}.$$

The assumptions (B) and (C) yields the result.

We conclude by following the same lines as in the proof of the Corollary 1.

Corollary 3 *If* (σ, F) *verifie the following condition:*

$$|\sigma(x)-\sigma(y)|^2+\int_{|z|<1}(F(x,z)-F(y,z))^2\lambda(dz)\le|f(x)-f(y)| \text{ for all } x, y$$

where $f:\mathbb{R}\to\mathbb{R}$ *is of bounded variation, and there exists* $\varepsilon>0$ *such that*

$$\sigma>\varepsilon \text{ or } F>\varepsilon$$

and if $x\to x+F(x,z)$ *is nondecreasing in a neighborhood of* 0, $\lambda(dz)$ *a.e. then* (σ, F) *verify* $(\mathscr{L}\mathscr{T})$ *condition.*

Proof The proof of the Corollary 3 is similar to what we have seen previously, this is due to the first hypothesis for σ and F which implies the (LT) condition and the hypothesis for F is identical to that of the Corollary 2.

Proposition 2 *If* σ *satisfies (LT) condition and* F *verifies the following condition:*

$$\int|F(x,z)-F(y,z)|\,\lambda(dz)\le c\,|x-y| \text{ for all } x, y \tag{11}$$

then the solution to (1) *is pathwise unique.*

Proof We use the Tanaka formula. Let X^1 and X^2 be two solutions to (1), then

$$\begin{aligned}|X_t^1-X_t^2|=&\int_0^t \operatorname{sign}(X_{s-}^1-X_{s-}^2)d(X_s^1-X_s^2)\\&+\sum_{0<s\le t}\{|X_s^1-X_s^2|-|X_{s-}^1-X_{s-}^2|\\&-\operatorname{sign}(X_{s-}^1-X_{s-}^2)\Delta(X_s^1-X_s^2)\}.\end{aligned}\tag{12}$$

Now, we have

$$\begin{aligned}d(X_t^1-X_t^2)=&(\sigma(X_t^1)-\sigma(X_t^2))dW_t+(b(X_t^1)-b(X_t^2))dt\\&+\int(F(X_{t-}^1,z)-F(X_{t-}^2,z))(\mu-\nu)(dt,dz).\end{aligned}\tag{13}$$

If we substitute (13) in (12), we obtain

$$
\begin{aligned}
|X_t^1 - X_t^2| &= \int_0^t \operatorname{sign}(X_{s-}^1 - X_{s-}^2)(\sigma(X_s^1) - \sigma(X_s^2))dW_s \\
&+ \int_0^t \operatorname{sign}(X_{s-}^1 - X_{s-}^2)(b(X_s^1) - b(X_s^2))ds \\
&+ \int_0^t \operatorname{sign}(X_{s-}^1 - X_{s-}^2) \int (F(X_{s-}^1, z) - F(X_{s-}^2, z))(\mu - \nu)(ds, dz) \\
&+ \sum_{0<s\le t} \left\{ |X_s^1 - X_s^2| - |X_{s-}^1 - X_{s-}^2| - \operatorname{sign}(X_{s-}^1 - X_{s-}^2)\Delta(X_s^1 - X_s^2) \right\}.
\end{aligned}
$$

The first and the third terms on the right hand side are martingales and all terms on the right hand side are integrable.
We can prove, if we note $I_2(t)$ the second term on the right hand side and using the fact that b is Lipschitz, that

$$
|I_2(t)| \le c \int_0^t |X_s^1 - X_s^2| \, ds. \tag{14}
$$

We treat the fourth term, denoted $I_4(t)$, by the following calculation:

$$
\begin{aligned}
|I_4(t)| &\le \sum_{0<s\le t} \left| |X_s^1 - X_s^2| - |X_{s-}^1 - X_{s-}^2| \right| + |\Delta(X_s^1 - X_s^2)| \\
&\le 2 \int_0^t \int |F(X_{s-}^1, z) - F(X_{s-}^2, z)| \, \mu(ds, dz) \\
&= 2 \int_0^t \int |F(X_{s-}^1, z) - F(X_{s-}^2, z)| \, (\mu - \nu)(ds, dz) \\
&+ 2 \int_0^t \int |F(X_{s-}^1, z) - F(X_{s-}^2, z)| \, \nu(ds, dz).
\end{aligned}
$$

Taking expectation in the two sides and using the martingales property and (11), we obtain

$$
\mathbb{E}(|I_4(t)|) \le c\mathbb{E}\left(\int_0^t ds \int |F(X_{s-}^1, z) - F(X_{s-}^2, z)| \, \lambda(dz) \right)
$$

and hence, we have

$$
\mathbb{E}(|I_4(t)|) \le c \int_0^t \mathbb{E}\left(|X_s^1 - X_s^2|\right) ds. \tag{15}
$$

Hence the result.

4 Others Results

Theorem 2 *If (σ, F) verify $(\mathscr{L}\mathscr{T})$ condition and b is measurable and bounded. Then the uniqueness in the sense of probability law implies the pathwise uniqueness.*

Proof Let X^1 and X^2 be two solutions of Eq. (1) with the same initial condition. We will prove that $Y = X^1 \vee X^2$ and $Z = X^1 \wedge X^2$ are solutions of (1). We obtain by using Tanaka's formula and $(\mathscr{L}\mathscr{T})$ condition,

$$(X_t^2 - X_t^1)^+ = \int_{0^+}^t \mathbf{1}_{(X_{s-}^2 - X_{s-}^1 > 0)} d(X_s^2 - X_s^1)$$

by using the fact that $X^1 \vee X^2 = X^1 + (X^2 - X^1)^+$ we obtain

$$\begin{aligned} X_t^1 \vee X_t^2 &= \int_0^t \Big[\mathbf{1}_{(X_{s-}^2 - X_{s-}^1 > 0)}(\sigma(X_s^2) - \sigma(X_s^1)) + \sigma(X_s^1)\Big] dW_s \\ &\quad + \int_0^t \Big[\mathbf{1}_{(X_{s-}^2 - X_{s-}^1 > 0)}(b(X_s^2) - b(X_s^1)) + b(X_s^1)\Big] ds \\ &\quad + \int_{0^+}^t \int \Big[\mathbf{1}_{(X_{s-}^2 - X_{s-}^1 > 0)}(F(X_{s-}^2, z) - F(X_{s-}^1, z)) \\ &\qquad + F(X_{s-}^1, z)\Big] (\mu - \nu)(ds, dz) \\ &= \int_0^t \sigma(X_s^1 \vee X_s^2) dW_s + \int_0^t b(X_s^1 \vee X_s^2) ds \\ &\quad + \int_{0^+}^t \int F((X^1 \vee X^2)_{s-}, z)(\mu - \nu)(ds, dz). \end{aligned}$$

Then Y is a solution of (1).
We have on the other hand,

$$(X_t^1 - X_t^2)^+ = \int_{0^+}^t \mathbf{1}_{(X_{s-}^1 - X_{s-}^2 > 0)} d(X_s^1 - X_s^2).$$

By using the fact that $X^1 \wedge X^2 = X^1 - (X^1 - X^2)^+$ we obtain by the same way that

$$\begin{aligned} X_t^1 \wedge X_t^2 &= \int_0^t \sigma(X_s^1 \wedge X_s^2) dW_s + \int_0^t b(X_s^1 \wedge X_s^2) ds \\ &\quad + \int_{0^+}^t \int F((X^1 \wedge X^2)_{s-}, z)(\mu - \nu)(ds, dz). \end{aligned}$$

Then Z is a solution of (1).

Finally, we have for all $t \geq 0$

$$\mathbb{E}\left[|X_t^1 - X_t^2|\right] = \mathbb{E}\left[X_t^1 \vee X_t^2\right] - \mathbb{E}\left[X_t^1 \wedge X_t^2\right],$$

and by using the uniqueness in the sense of probability law, we obtain

$$\mathbb{E}\left[|X_t^1 - X_t^2|\right] = 0.$$

Since X^1 and X^2 are càdlàg, hence $X_t^1 = X_t^2$ for all $t \geq 0$ a.s.

This allows us to give a generalization of Bass's result [4], we have the following:

Theorem 3 *If* (σ, F) *verify* $(\mathscr{L}\mathscr{T})$ *condition and* b *is measurable, bounded and moreover we suppose that:*

- σ *is bounded, continuous and strictly positive.*
- $x \to \int_A \frac{|z|^2}{1+|z|^2} F(x, z)\lambda(dz)$ *is bounded and continuous for each* $A \subset \mathbb{R} - \{0\}$.

Then there exists a solution to (1) *that is pathwise unique.*

Proof By Bass's result [4], the hypothesis of theorem entail the existence of a unique solution to the martingale problem associated to the SDE (1). By using the equivalence between martingale problem- stochastic differential equation, this implies existence of a unique weak solution to the Eq. (1). Hence, by the $(\mathscr{L}\mathscr{T})$ condition combined with last theorem, we get the desired result.

If we set $F(x, z) = \frac{1}{|z|^{1+\alpha(x)}}$ and $\sigma = b = 0$ in the Eq. (1), we obtain the stable-like process with the operator

$$\mathscr{L} f(x) = \int \left[f(x+z) - f(x) - \mathbf{1}_{(|z|\leq 1)} f'(x)z\right] \frac{1}{|z|^{1+\alpha(x)}} dz.$$

We have the following proposition

Proposition 3 *If the function* α *is Dini continuous, bounded above by a constant less than 2 and bounded bellow by a constant greater than 0 and is increasing, then we have pathwise uniqueness of the solution of the stochastic differential equation driven by stable-like process associated to F.*

The proof is a consequence of Bass's result ([4], p. 13).

Theorem 4 *Suppose that for* $i = 1, 2$, X^i *satisfies:*

$$dX_t^i = \sigma(X_t^i)dW_t + b_i(X_t^i)dt + \int F(X_{t-}^i, z)(\mu - \nu)(dt, dz), \qquad (16)$$

where σ, F, b_1, b_2, *are bounded measurable functions. Assume that:*

- (σ, F) *verify* $(\mathscr{L}\mathscr{T})$ *condition.*

- *One of the two functions b_1, b_2 is Lipschitz.*

Assume further that:

1. $b_1 \le b_2$.
2. $X_0^1 \le X_0^2$.

Then: $X_t^1 \le X_t^2$ for all t a.s.

Proof Let $X_i, i = 1, 2$ two solutions of equations (16). By Tanaka formula we obtain

$$\begin{aligned}(X_t^1 - X_t^2)^+ &= \int_0^t \mathbf{1}_{(X_{s-}^1 - X_{s-}^2 > 0)} d(X_s^1 - X_s^2) \\ &+ \sum_{0<s\le t} \mathbf{1}_{(X_{s-}^1 - X_{s-}^2 > 0)} \left(X_s^1 - X_s^2\right)^- \\ &+ \sum_{0<s\le t} \mathbf{1}_{(X_{s-}^1 - X_{s-}^2 \le 0)} \left(X_s^1 - X_s^2\right)^+ \\ &+ \frac{1}{2} L_t^0(X^1 - X^2). \end{aligned} \tag{17}$$

As (σ, F) verify $(\mathscr{L}\mathscr{T})$ condition, the second, the third and the fourth terms in the right hand sides are zero. Using the same argument as in Proposition 2, we find

$$\mathbb{E}\left[\left(X_t^1 - X_t^2\right)^+\right] \le c \int_0^t \mathbb{E}\left[\left(X_s^1 - X_s^2\right)^+\right] ds$$

which implies that $X_t^1 \le X_t^2$ for all $t \ge 0$ a.s. (since X^1 and X^2 are càdlàg).

Theorem 5 *Suppose that for $i = 1, 2$, X^i satisfies:*

$$dX_t^i = \sigma(X_t^i) dW_t + b_i(X_t^i) dt + \int F_i(X_{t-}^i, z)(\mu - \nu)(dt, dz) \tag{18}$$

where $\sigma, F_1, F_2, b_1, b_2$, are bounded measurable functions. Assume that:

- *σ verify (LT) condition.*
- *One of the two functions b_1, b_2 is Lipschitz.*

Assume further that:

1. $b_1 \le b_2$.
2. $X_0^1 \le X_0^2$.
3. $x_1 + F_1(x_1, z) \le x_2 + F_2(x_2, z)$ *for all* $x_1 \le x_2$.

Then: $X_t^1 \le X_t^2$ for all t a.s.

Proof We apply the same method as before in the Theorem 4, the second and the third terms in the right hand side of (17) is zero because of the assumption 3 and the fourth term is zero because the σ verify (LT) condition. The end of the proof is identical to that of the Theorem 4.

Remark 1 We can see that if $F_i, i = 1, 2$ satisfy the hypothesis of Corollary 2 (or 3), then we can replace the hypothesis 3. by

$$F_1(x_1, z) \le F_2(x_2, z), \forall x_1 \le x_2.$$

Remark 2 If we take $F_1 = F_2$ in the Theorem 2, the assumption of the Corollary 1 (or 2) on F is enough for the conclusion of the Theorem 4.

Remark 3 We can remark that the condition 3 in the Theorem 5 ensures that the paths of X_i do not cross at jump times: if (s, z) is an atom of μ and if $X^1_{s-} = x \le X^2_{s-} = y$, then

$$X^1_s = x + F_1(x, z) \le y + F_2(y, z) = X^2_s$$

and this condition is necessary for comparison theorem.

Remark 4 We can remark that the Theorem 5 generalizes the result of Peng and Zhu [16] namely Theorem 3.1.

Remark 5 Our approach based on local time technic can be used for more general equation of type:

$$\begin{aligned} X_t = X_0 &+ \int_0^t \sigma(X_{s-}, u)\ W(ds, du) + \int_0^t b(X_{s-})\ ds \\ &+ \int_0^t \int_{\{|u|\le 1\}} g_0(X_{s-}, u)\widetilde{N}_0(ds, du) \\ &+ \int_0^t \int_{\{|u|\ge 1\}} g_1(X_{s-}, u) N_1(ds, du) \end{aligned}$$

where

- $\{W(ds, du)\}$ the white noise with intensity $ds\pi(dz)$, with π is a $\sigma-$finite measure on $\mathbb{R}$.
- $N_1(ds, du)$ and $\widetilde{N}_0(ds, du)$ denote the Poisson random measures on $[0, \infty) \times [-1, 1]$, $[0, \infty) \times [-1, 1]^c$ respectively, defined on same probability space and are independent each of other.
- $\widetilde{N}_0(ds, du)$ denote the compensated measure of $N_0(ds, du)$.

This SDE was recently studied by Donald A. Dawson and Zenghu Li [6] and treated by Ouknine [15] in continuous case, it will be the subject of another paper.

References

1. Barndorff-Nielsen, O.E.: Processes of normal inverse Gaussian type. Finance Stochast. **2**, 41–68 (1998)
2. Barlow, M.T., Perkins, E.: SDE's singular increasing process. Stochastics **12**, 229–249 (1984)
3. Bass, R.: Stochastic differential equations driven by symmetric stable processes. In: Séminaire de probabilités, XXXVI, Lecture Notes in Math., Vol. 1801, pp. 302–313. Springer, Berlin (2003)
4. Bass, R.: Stochastic differential equations with jumps. Probab. Surv. **1**, 1–19 (2004)
5. Cont, R., Tankov, P.: Financial modeling with jump processes, Chapman and Hall/CRC (2003)
6. Dawson, D.A., Li, Z.: Stochastic equations, flows and measure-valued processes. Ann. Probab. **40**(2), 813–857 (2012)
7. Engelbert, H.J., Schmidt, W.: Strong Markov continous local martingales and solutions of one-Dimensional Stochastic differential equations. Part I. Math. Nachr. **143**, 167–184 (1989)
8. Fu, Z., Li, Z.: Stochastic equations of non-negative processes with jumps. Stochastic Process. Appl. **120**(3), 306–330 (2010)
9. Ikeda, N., Watanabe, S.: Stochastic differential equations and diffusion processes. North-Holland Mathematical Library, vol. 24, 2nd edn. North-Holland Publishing Co., Amsterdam (1989)
10. Le Gall, J. F.: One-dimensional stochastic differential equations involving the local cfftimes of unknown processes. In: Proceedings Stochastic Analysis and Applications (Swansea, 1983), Lecture Notes in Math., Vol. 1095, pp. 51–82. Springer, Berlin (1984)
11. Ornstein, L.S., Uhlenbeck, G.E.: On the theory of Brownain motion. Physical Review **36**, 823–841 (1930)
12. Ouknine, Y.: Généralisation d'un lemme de S. Nakao et applications. Stochastics **23**(2), 149–157 (1988)
13. Ouknine, Y.: Fonctions de semimartingales et applications aux équations différentielles stochastiques. Stochast. Stochast. Rep. **28**, 115–122 (1989)
14. Ouknine, Y.: Quelques identités sur les temps locaux et unicité des solutions d'équations differentielles stochastiques avec reflection. Stochastic Process. Appl. **48**(2), 335–340 (1993)
15. Ouknine, Y.: Unicité trajectorielle des equations différentielles stochastiques avec temps local. Probab. Math. Stat. **19**(1), 55–62 (1999)
16. Peng, S., Zhu, X.: Necessary and sufficient condition for comparison theorem of 1-dimensional stochastic differential equations. Stochastic Process. Appl. **116**, 370–380 (2006)
17. Protter, P.: Stochastic integration and differential equations, 2nd edn. Springer, Berlin Heidelberg New York (2003)
18. Rutkowski, M.: Stochastic differential with singular drift. Stat. Probabil. Lett. **10**(3), 225–229 (1990)
19. Samuelson, P.A.: Rational theory of warrant pricing. Ind. Management Rev. **6**, 1339 (1965)
20. Skorokhod, A.V.: Studies in the theory of random processes. Addison-Wesley Publishing Co., Inc, Reading, Mass (1965)

BSDE Approach for Dynkin Game and American Game Option

El Hassan Essaky and M. Hassani

Abstract Consider a Dynkin game with payoff

$$J(\lambda, \sigma) = F\left[U_\lambda 1_{\{\lambda<\sigma\}} + L_\sigma 1_{\{\lambda>\sigma\}} + Q_\sigma 1_{\{\sigma=\lambda<T\}} + \xi 1_{\{\sigma=\lambda=T\}} \right],$$

where $F : \mathbb{R} \longrightarrow \mathbb{R}$ is a continuous nondecreasing function and λ, σ are stopping times valued in $[0, T]$. We show the existence of a value as well as a saddle-point for this game using the theory of BSDE with double reflecting barriers. An American game option pricing problem is also discussed.

Keywords Backward stochastic differential equations with double reflecting barriers · Dynkin game · Saddle-point · American game option

Mathematical Subject Classification 2010 60H10 · 60H20 · 60H30

1 Introduction

Stochastic game was first introduced by Dynkin and Yushkevich [6] and later studied, in different contexts, by several authors, including Neveu [18], Bensoussan and Friedman [2], Bismut [3], Morimoto [17], Alario-Nazaret, Lepeltier and Marchal [1], Lepeltier and Maingueneau [16], Cvitanic and Karatzas [4], Touzi and Vieille [19] and others, such stochastic games are known as Dynkin games.

E.H. Essaky (✉) · M. Hassani
Département de Mathématiques et d'Informatique,
Laboratoire de Modélisation et Combinatoire, Faculté Poly-disciplinaire,
Université Cadi Ayyad, B.P. 4162, Safi, Morocco
e-mail: essaky@uca.ma

M. Eddahbi et al. (eds.), *Statistical Methods and Applications in Insurance and Finance*, Springer Proceedings in Mathematics & Statistics 158,
DOI 10.1007/978-3-319-30417-5_9

Considerable attention has been devoted to studying the association between backward stochastic differential equations (BSDEs for short) and stochastic differential games. Among others, Cvitanic and Karatzas showed in [4] existence and uniqueness of the solution to the BSDE with double reflecting barriers, and associated their equation to a stochastic games. Hamadène [9] and Hamadène and Hassani [11] studied the mixed zero-sum stochastic differential game problem using the notion of a local solution of BSDEs with double reflecting barriers. Hamadène and Lepeltier [10] added controls to the Dynkin game studied by Cvitanic and Karatzas in [4]. Karatzas and Li [14] studied a non-zero-sum game with features of both stochastic control and optimal stopping, for a process of diffusion type via the BSDE approach. Dumitrescu et al. [5] introduced a generalized Dynkin game problem associated with a BSDE with jumps.

Consider the Dynkin game, associated with processes L, U, ξ and Q, with payoff:

$$J(\lambda, \sigma) = F\Big[U_\lambda 1_{\{\lambda<\sigma\}} + L_\sigma 1_{\{\lambda>\sigma\}} + Q_\sigma 1_{\{\sigma=\lambda<T\}} + \xi 1_{\{\sigma=\lambda=T\}}\Big],$$

where $F : \mathbb{R} \longrightarrow \mathbb{R}$ is a continuous nondecreasing function and λ, σ are stopping times valued in $[0, T]$. In the direction of connection between BSDE with two reflecting barriers and Dynkin games, in order to prove that this game has a saddle point, which is a pair of stopping (λ^*, σ^*) such that for any stopping times λ and σ one has

$$\mathbb{E}\Big(J(\lambda^*, \sigma)\Big) \le \mathbb{E}\Big(J(\lambda^*, \sigma^*)\Big) \le \mathbb{E}\Big(J(\lambda, \sigma^*)\Big),$$

all the works [4, 9–11, 14] have considered the case of bounded or square integrable processes $F(\xi)$, $F(Q)$, $F(L)$ and $F(U)$. Moreover, they have assumed that the barriers $F(L)$ and $F(U)$ have to satisfy one of the conditions:

1. The so-called Mokobodski condition which turns out into the existence of a difference of nonnegative supermartingales between $F(L)$ and $F(U)$.
2. The complete separation i.e. $F(L) < F(U)$.

One of the main objective of this work is to weaken the assumptions assumed on the data $F(\xi)$, $F(Q)$, $F(L)$ and $F(U)$ in the case of association between BSDE with two reflecting barriers and Dynkin games. Yet, checking Mokobodski's condition appears as a difficult question. So, instead of assuming the Mokobodski's condition on the barriers $F(L)$ and $F(U)$, we suppose only that there exists a semimartingale between them. It should be also noted here that if the barriers are completely separated this implies that there exists a semimartingale between them (see [8]). Actually, if we assume the following conditions:

1. There exists a semimartingale between L and U and for every semimartingale S such that $L \le S \le U$, $F(S)$ is a also a semimartingale.
2. $\mathbb{E}[F(L_\sigma)^-] < +\infty$, for all stopping time $0 \le \sigma \le T$, where $F(L)^- = \sup(-F(L), 0)$.

3. $\liminf_{r\to+\infty} r\, P\left[\sup_{s\le T} F(U_s)^+ > r\right] = 0$, where $F(U)^+ = \sup(F(U), 0)$.
4. $\liminf_{r\to+\infty} r\, P\left[\sup_{s\le T} F(L_s)^- > r\right] = 0$,

then the pair of stopping times (λ^*, σ^*) defined by

$$\lambda^* = \inf\{s \ge 0 \;:\; Y_s = F(U_s)\} \wedge T \quad \text{and} \quad \sigma^* = \inf\{s \ge 0 \;:\; Y_s = F(L_s)\} \wedge T,$$

is a saddle-point for the game, where Y is the solution of the following BSDE with double reflecting barriers $F(L)$ and $F(U)$ (see Definition 2):

$$\begin{cases} (i) \;\; Y_t = F(\xi) + \int_t^T dK_s^+ - \int_t^T dK_s^- - \int_t^T Z_s dB_s \,, t \le T, \\ (ii) \;\; Y \text{ between } F(L) \text{ and } F(U), \; i.e. \; \forall t \le T, \;\; F(L_t) \le Y_t \le F(U_t), \\ (iii) \;\; \text{the Skorohod conditions hold:} \\ \qquad \int_0^T (Y_t - F(L_t))dK_t^+ = \int_0^T (F(U_t) - Y_t)dK_t^- = 0, \;\; \text{a.s..} \end{cases}$$

We should mention here that if $F(L)$ and $F(U)$ are L^1—integrable, i.e. $\mathbb{E}\sup_{t\le T}(|F(U_t)| + |F(L_t)|) < +\infty$, then the above assumptions 2–4 are satisfied and then the Dynkin game has a saddle point. This corresponds to the main assumption assumed in the general context of Dynkin games.

An American option is a contract which enables its buyer (holder) to exercise it at any time up to the maturity. An American game option gives additionally the right to the option seller (writer, issuer) to cancel it early paying for this a prescribed penalty. American game option was first introduced by Kifer [15] and studied later by several authors, see for example Hamadène [9], Hamadène and Zhang [13] and the references therein. The second aim of this work is to prove, under weaker conditions than the square integrability assumed on the data in [9], that the value of the option at any time $t \in [0, T]$ is given by $e^{rt} Y_t$, where Y is the solution of some BSDE with two reflecting barriers. Moreover, we also show that a hedge after t, against the game option, exists.

2 Preliminaries

2.1 Notations and Assumptions

Let $(\Omega, \mathscr{F}, (\mathscr{F}_t)_{t\le T}, P)$ be a stochastic basis on which is defined a Brownian motion $(B_t)_{t\le T}$ such that $(\mathscr{F}_t)_{t\le T}$ is the natural filtration of $(B_t)_{t\le T}$ and $\mathscr{F}_0$ contains all P-null sets of $\mathscr{F}$. Note that $(\mathscr{F}_t)_{t\le T}$ satisfies the usual conditions, i.e. it is right continuous and complete.

Let us now introduce the following notations:

- $\mathscr{P}$ the sigma algebra of $\mathscr{F}_t$-progressively measurable sets on $\Omega \times [0, T]$.
- $\mathscr{C}$ the set of $\mathbb{R}$-valued $\mathscr{P}$-measurable continuous processes $(Y_t)_{t\leq T}$.
- $\mathscr{L}^{2,d}$ the set of $\mathbb{R}^d$-valued and $\mathscr{P}$-measurable processes $(Z_t)_{t\leq T}$ such that

$$\int_0^T |Z_s|^2 ds < \infty, \ P - a.s.$$

- $\mathscr{K}$ the set of $\mathscr{P}$-measurable continuous nondecreasing processes $(K_t)_{t\leq T}$ such that $K_0 = 0$ and $K_T < +\infty$, P– a.s.

Throughout the paper, we introduce the following data:

- $L := \{L_t, \ 0 \leq t \leq T\}$ and $U := \{U_t, \ 0 \leq t \leq T\}$ are two real valued barriers which are $\mathscr{P}$-measurable and continuous processes such that $L_t \leq U_t, \ \forall t \in [0, T]$.
- Q be a process such that, $\forall t \in [0, T]$ $L_t \leq Q_t \leq U_t, \ P - a.s.$
- ξ is an $\mathscr{F}_T$-measurable one dimensional random variable such that

$$L_T \leq \xi \leq U_T.$$

- $F : \mathbb{R} \longrightarrow \mathbb{R}$ is a continuous nondecreasing function.

We assume the following assumptions:

(**A.1**) There exists a continuous semimartingale $S_. = S_0 + V^+_. - V^-_. + \int_0^. \alpha_s dB_s$, with $S_0 \in \mathbb{R}$, $V^+, V^- \in \mathscr{K}$ and $\alpha \in \mathscr{L}^{2,d}$, such that

$$L_t \leq S_t \leq U_t, \ \ \forall t \in [0, T].$$

(**A.2**) For every semimartingale S such that $L \leq S \leq U$, $F(S)$ is a also a semimartingale.

2.2 *Existence of Solution for BSDE with Double Reflecting Barriers*

In view of clarifying this issue, we recall some results concerning BSDEs with double reflecting barriers with two continuous barriers (see Essaky and Hassani [8] for more details). Let us recall first the following definition of two singular measures.

Definition 1 Let K^1 and K^2 be two processes in $\mathscr{K}$. We say that:

K^1 and K^2 are singular if and only if there exists a set $D \in \mathscr{P}$ such that

$$\mathbb{E}\int_0^T 1_D(s,\omega)dK_s^1(\omega) = \mathbb{E}\int_0^T 1_{D^c}(s,\omega)dK_s^2(\omega) = 0.$$

This is denoted by $dK^1 \perp dK^2$.

Let us now introduce the definition of a BSDE with double reflecting obstacles L and U.

Definition 2 1. We call $(Y, Z, K^+, K^-) := (Y_t, Z_t, K_t^+, K_t^-)_{t\leq T}$ a solution of the GBSDE with two reflecting barriers L and U associated with a terminal value ξ if the following hold:

$$\begin{cases} (i)\ \ Y_t = \xi + \int_t^T dK_s^+ - \int_t^T dK_s^- - \int_t^T Z_s dB_s\,, t \leq T, \\ (ii)\ \ Y \text{ between } L \text{ and } U,\ i.e.\ \forall t \leq T,\ L_t \leq Y_t \leq U_t, \\ (iii)\ \ \text{the Skorohod conditions hold:} \\ \qquad \int_0^T (Y_t - L_t)dK_t^+ = \int_0^T (U_t - Y_t)dK_t^- = 0, \text{ a.s.}, \\ (iv)\ Y \in \mathscr{C} \quad K^+, K^- \in \mathscr{K} \quad Z \in \mathscr{L}^{2,d}, \\ (v)\ \ dK^+ \perp dK^-. \end{cases} \tag{1}$$

2. We say that the BSDE (1) has a maximal (resp. minimal) solution (Y, Z, K^+, K^-) if for any other solution (Y', Z', K'^+, K'^-) of (1) we have for all $t \leq T$, $Y_t' \leq Y_t$, P − a.s. (resp. $Y_t' \geq Y_t$, P − a.s.).

The following theorem has already been proved in [8].

Theorem 1 *Let assumption* (**A.1**) *holds true. Then there exists a maximal (resp. minimal) solution for BSDE with double reflecting barriers* (1).

3 Dynkin Game

Our purpose in this section is to show that the existence of a solution (Y, Z, K^+, K^-) to the BSDE (1) implies that Y is the value of a certain stochastic game of stopping.

Consider the payoff

$$J(\lambda, \sigma) = F\Big(U_\lambda 1_{\{\lambda<\sigma\}} + L_\sigma 1_{\{\lambda>\sigma\}} + Q_\sigma 1_{\{\sigma=\lambda<T\}} + \xi 1_{\{\sigma=\lambda=T\}}\Big).$$

The setting of our problem of Dynkin game is the following. There are two players labeled player 1 and player 2. Player 1 chooses the stopping time λ, player 2 chooses the stopping time σ, and $J(\lambda, \sigma)$ represents the amount paid by player 1 to player 2. It is the conditional expectation $\mathbb{E}\Big(J(\lambda, \sigma)\Big)$ of this random payoff that player 1 tries to minimize and player 2 tries to maximize. The game stops when one player decides to stop, that is, at the stopping time $\lambda \wedge \sigma$ before time T, the payoff is then equals

$$J(\lambda, \sigma) = \begin{cases} F(U_\lambda) \text{ if player 1 stops the game first} \\ F(L_\sigma) \text{ if player 2 stops the game first} \\ F(Q_\sigma) \text{ if players stop the game simultaneously before time T} \\ F(\xi) \text{ if neither have exercised until the expiry time T.} \end{cases}$$

It is then natural to define the lower and upper values of the game:

$$\underline{V} := \sup_{\sigma \in \mathcal{M}_{t,T}} \inf_{\lambda \in \mathcal{M}_{t,T}} \mathbb{E}\Big[J(\lambda, \sigma)\Big] \leq \overline{V} := \inf_{\lambda \in \mathcal{M}_{t,T}} \sup_{\sigma \in \mathcal{M}_{t,T}} \mathbb{E}\Big[J(\lambda, \sigma)\Big],$$

where $\mathcal{M}_{t,T}$ is the set of stopping times valued between t and T. If it happens that $\underline{V} = \overline{V}$, then the above Dynkin game is said to have a value. A pair $(\lambda_0^*, \sigma_0^*)$ is called a saddle point if

$$\mathbb{E}\Big(J(\lambda_0^*, \sigma)\Big) \leq \mathbb{E}\Big(J(\lambda_0^*, \sigma_0^*)\Big) \leq \mathbb{E}\Big(J(\lambda, \sigma_0^*)\Big).$$

Our objective is to show the existence of a saddle-point for the game and to characterize it. This implies that this game has a value.

Let assumptions (**A.1**) and (**A.2**) hold true. Let (Y, Z, K^+, K^-) be the solution, which is exists according to Theorem 1, of the following BSDE with double reflecting barriers:

$$\begin{cases} (i) \;\; Y_t = F(\xi) + \int_t^T dK_s^+ - \int_t^T dK_s^- - \int_t^T Z_s dB_s, t \leq T, \\ (ii) \;\; \forall t \leq T, \;\; F(L_t) \leq Y_t \leq F(U_t), \\ \qquad \int_0^T (Y_t - F(L_t))dK_t^+ = \int_0^T (F(U_t) - Y_t)dK_t^- = 0, \text{ a.s.,} \\ (iv) \;\; Y \in \mathscr{C} \quad K^+, K^- \in \mathscr{K} \quad Z \in \mathscr{L}^{2,d}, \\ (v) \;\; dK^+ \perp dK^-. \end{cases} \tag{2}$$

Let λ_t^* and σ_t^* be the stopping times defined as follows:

$$\lambda_t^* = \inf\{s \geq t \;:\; Y_s = F(U_s)\} \wedge T \quad \text{and} \quad \sigma_t^* = \inf\{s \geq t \;:\; Y_s = F(L_s)\} \wedge T.$$

The main result of this section is the following.

Theorem 2 *Assume the following assumptions:*

1. $\mathbb{E}F(L_\sigma)^- < +\infty$, *for all stopping time* $0 \le \sigma \le T$, *where* $F(L)^- = \sup(-F(L), 0)$.
2. $\liminf_{r\to+\infty} r\, P\left[\sup_{s\le T} F(U_s)^+ > r\right] = 0$, *where* $F(U)^+ = \sup(F(U), 0)$.
3. $\liminf_{r\to+\infty} r\, P\left[\sup_{s\le T} F(L_s)^- > r\right] = 0$.

Then

$$\begin{aligned} Y_t &= \mathbb{E}\Big[J(\lambda_t^*, \sigma_t^*) \mid \mathcal{F}_t\Big] \\ &= \sup_{\sigma\in\mathcal{M}_{t,T}} \mathbb{E}\Big[J(\lambda_t^*, \sigma) \mid \mathcal{F}_t\Big] = \inf_{\lambda\in\mathcal{M}_{t,T}} \mathbb{E}\Big[J(\lambda, \sigma_t^*) \mid \mathcal{F}_t\Big] \\ &= \inf_{\lambda\in\mathcal{M}_{t,T}} \sup_{\sigma\in\mathcal{M}_{t,T}} \mathbb{E}\Big[J(\lambda, \sigma) \mid \mathcal{F}_t\Big] = \sup_{\sigma\in\mathcal{M}_{t,T}} \inf_{\lambda\in\mathcal{T}_t} \mathbb{E}\Big[J(\lambda, \sigma) \mid \mathcal{F}_t\Big], \end{aligned} \tag{3}$$

where $\mathcal{M}_{t,T}$ *is the set of stopping times valued between* t *and* T. Y_0 *can be interpreted as the value of the game and* $(\lambda_0^*, \sigma_0^*)$ *as the fair strategy for the two players (or a saddle point for the game).*

Proof Let $(a_n^+)_n$ and $(a_n^-)_n$ be two nondecreasing sequences such that

$$\liminf_{n\to+\infty} a_n^+ P\left[\sup_{s\le T} F(U_s)^+ > a_n^+\right] = 0, \quad \liminf_{n\to+\infty} a_n^- P\left[\sup_{s\le T} F(L_s)^- > a_n^-\right] = 0. \tag{4}$$

Let also $(\alpha_i)_{i\ge0}$ and $(\upsilon_i^\pm)_{i\ge0}$ be families of stopping times defined by

$$\alpha_i = \inf\{s \ge t : \int_t^s |\, Z_r \,|^2\, dr \ge i\} \wedge T, \quad \upsilon_i^\pm = \inf\{s \ge t : Y_s^\pm > a_i^\pm\} \wedge T.$$

It follows from Eq. (2) that for every stopping time $\sigma \in \mathcal{M}_{t,T}$

$$\begin{aligned} Y_t =& Y_{\lambda_t^*\wedge\sigma\wedge\alpha_i\wedge\upsilon_n^+\wedge\upsilon_m^-} + \int_t^{\lambda_t^*\wedge\sigma\wedge\alpha_i\wedge\upsilon_n^+\wedge\upsilon_m^-} dK_s^+ - \underbrace{\int_t^{\lambda_t^*\wedge\sigma\wedge\alpha_i\wedge\upsilon_n^+\wedge\upsilon_m^-} dK_s^-}_{=0} \\ & - \int_t^{\lambda_t^*\wedge\sigma\wedge\alpha_i\wedge\upsilon_n^+\wedge\upsilon_m^-} Z_s dB_s. \end{aligned}$$

Then for every stopping time $\sigma \in \mathcal{M}_{t,T}$

$$
\begin{aligned}
Y_t &\geq \mathbb{E}\left(Y_{\lambda_t^*\wedge\sigma\wedge\alpha_i\wedge\upsilon_n^+\wedge\upsilon_m^-} \mid \mathscr{F}_t\right)\\
&= \mathbb{E}\left(Y^+_{\lambda_t^*\wedge\sigma\wedge\alpha_i\wedge\upsilon_n^+\wedge\upsilon_m^-} \mid \mathscr{F}_t\right) - \mathbb{E}\left(Y^-_{\lambda_t^*\wedge\sigma\wedge\alpha_i\wedge\upsilon_n^+\wedge\upsilon_m^-} \mid \mathscr{F}_t\right).
\end{aligned}
$$

In view of passing to the limit on i and n respectively and using Fatou's lemma for Y^+ and dominated convergence theorem for Y^- since it is bounded, we have

$$
Y_t \geq \mathbb{E}\left(Y^+_{\lambda_t^*\wedge\sigma\wedge\upsilon_m^-} \mid \mathscr{F}_t\right) - \mathbb{E}\left(Y^-_{\lambda_t^*\wedge\sigma\wedge\upsilon_m^-} \mid \mathscr{F}_t\right).
$$

Now taking the upper limit on m we get

$$
\begin{aligned}
Y_t &\geq \limsup_m \left[\mathbb{E}\left(Y^+_{\lambda_t^*\wedge\sigma\wedge\upsilon_m^-} \mid \mathscr{F}_t\right) - \mathbb{E}\left(Y^-_{\lambda_t^*\wedge\sigma\wedge\upsilon_m^-} \mid \mathscr{F}_t\right)\right]\\
&= \limsup_m \left[\mathbb{E}\left(Y^+_{\lambda_t^*\wedge\sigma} 1_{\lambda_t^*\wedge\sigma\leq\upsilon_m^-} \mid \mathscr{F}_t\right) + \mathbb{E}\left(Y^+_{\upsilon_m^-} 1_{\lambda_t^*\wedge\sigma>\upsilon_m^-} \mid \mathscr{F}_t\right)\right.\\
&\quad \left.-\mathbb{E}\left(Y^-_{\lambda_t^*\wedge\sigma\wedge\upsilon_m^-} \mid \mathscr{F}_t\right)\right]\\
&\geq \limsup_m \left[\mathbb{E}\left(Y^+_{\lambda_t^*\wedge\sigma} 1_{\lambda_t^*\wedge\sigma\leq\upsilon_m^-} \mid \mathscr{F}_t\right) - \mathbb{E}\left(Y^-_{\lambda_t^*\wedge\sigma\wedge\upsilon_m^-} \mid \mathscr{F}_t\right)\right]\\
&= \mathbb{E}\left(Y^+_{\lambda_t^*\wedge\sigma} \mid \mathscr{F}_t\right) - \liminf_m \mathbb{E}\left(Y^-_{\lambda_t^*\wedge\sigma\wedge\upsilon_m^-} \mid \mathscr{F}_t\right).
\end{aligned}
$$

In view of using the limit appearing in (4), we obtain

$$
\begin{aligned}
&\liminf_m \mathbb{E}\left(Y^-_{\lambda_t^*\wedge\sigma\wedge\upsilon_m^-} \mid \mathscr{F}_t\right)\\
&\leq \liminf_m \left[\mathbb{E}\left(Y^-_{\lambda_t^*\wedge\sigma} 1_{\lambda_t^*\wedge\sigma\leq\upsilon_m^-} \mid \mathscr{F}_t\right) + a_m^- \mathbb{E}\left(1_{\lambda_t^*\wedge\sigma>\upsilon_m^-} \mid \mathscr{F}_t\right)\right]\\
&= \mathbb{E}\left(Y^-_{\lambda_t^*\wedge\sigma} \mid \mathscr{F}_t\right) + \liminf_{m\to+\infty} a_m^- \mathbb{E}\left(1_{\lambda_t^*\wedge\sigma>\upsilon_m^-} \mid \mathscr{F}_t\right)\\
&\leq \mathbb{E}\left(Y^-_{\lambda_t^*\wedge\sigma} \mid \mathscr{F}_t\right) + \liminf_{m\to+\infty} a_m^- \mathbb{E}\left(1_{\{\sup_{s\leq T} F(L_s)^- > a_m^-\}} \mid \mathscr{F}_t\right)\\
&= \mathbb{E}\left(Y^-_{\lambda_t^*\wedge\sigma} \mid \mathscr{F}_t\right),
\end{aligned}
$$

it follows then that for all stopping time $\sigma \in \mathscr{M}_{t,T}$,

$$
\begin{aligned}
Y_t &\geq \mathbb{E}\Big(Y^+_{\lambda^*_t\wedge\sigma} \mid \mathscr{F}_t\Big) - \mathbb{E}\Big(Y^-_{\lambda^*_t\wedge\sigma} \mid \mathscr{F}_t\Big) = \mathbb{E}\Big(Y_{\lambda^*_t\wedge\sigma} \mid \mathscr{F}_t\Big)\\
&\geq \mathbb{E}\Big(F(U_{\lambda^*_t})1_{\{\lambda^*_t<\sigma\}} + F(L_\sigma)1_{\{\lambda^*_t>\sigma\}} + F(Q_\sigma)1_{\{\sigma=\lambda^*_t<T\}} + F(\xi)1_{\{\sigma=\lambda^*_t=T\}} \mid \mathscr{F}_t\Big)\\
&= \mathbb{E}\Big(J(\lambda^*_t,\sigma) \mid \mathscr{F}_t\Big).
\end{aligned}
\tag{5}
$$

Now it follows from Eq. (2) that for every stopping time $\lambda \in \mathscr{M}_{t,T}$

$$
\begin{aligned}
Y_t &\leq \mathbb{E}\Big(Y_{\lambda\wedge\sigma^*_t\wedge\alpha_i\wedge\upsilon^-_m\wedge\upsilon^+_n} \mid \mathscr{F}_t\Big)\\
&= \mathbb{E}\Big(Y^+_{\lambda\wedge\sigma^*_t\wedge\alpha_i\wedge\upsilon^-_m\wedge\upsilon^+_n} \mid \mathscr{F}_t\Big) - \mathbb{E}\Big(Y^-_{\lambda\wedge\sigma^*_t\wedge\alpha_i\wedge\upsilon^-_m\wedge\upsilon^+_n} \mid \mathscr{F}_t\Big).
\end{aligned}
$$

In view of passing to the limit on i and m respectively and using dominated convergence theorem for Y^+ since it is bounded, we have

$$
\begin{aligned}
&Y_t\\
&\leq \mathbb{E}\Big(Y^+_{\lambda\wedge\sigma^*_t\wedge\upsilon^+_n} \mid \mathscr{F}_t\Big) - \limsup_m \mathbb{E}\Big(Y^-_{\lambda\wedge\sigma^*_t\wedge\upsilon^-_m\wedge\upsilon^+_n} \mid \mathscr{F}_t\Big)\\
&= \mathbb{E}\Big(Y^+_{\lambda\wedge\sigma^*_t\wedge\upsilon^+_n} \mid \mathscr{F}_t\Big) - \limsup_m \Big[\mathbb{E}\Big(Y^-_{\lambda\wedge\sigma^*_t\wedge\upsilon^+_n} 1_{\lambda\wedge\sigma^*_t\wedge\upsilon^+_n\leq\upsilon^-_m} \mid \mathscr{F}_t\Big)\\
&\quad +\mathbb{E}\Big(Y^-_{\upsilon^-_m} 1_{\lambda\wedge\sigma^*_t\wedge\upsilon^+_n>\upsilon^-_m} \mid \mathscr{F}_t\Big)\Big]\\
&\leq \mathbb{E}\Big(Y^+_{\lambda\wedge\sigma^*_t\wedge\upsilon^+_n} \mid \mathscr{F}_t\Big) - \mathbb{E}\Big(Y^-_{\lambda\wedge\sigma^*_t\wedge\upsilon^+_n} \mid \mathscr{F}_t\Big) - \limsup_m \mathbb{E}\Big(Y^-_{\upsilon^-_m} 1_{\lambda\wedge\sigma^*_t\wedge\upsilon^+_n>\upsilon^-_m} \mid \mathscr{F}_t\Big)\\
&\leq \mathbb{E}\Big(Y^+_{\lambda\wedge\sigma^*_t\wedge\upsilon^+_n} \mid \mathscr{F}_t\Big) - \mathbb{E}\Big(Y^-_{\lambda\wedge\sigma^*_t\wedge\upsilon^+_n} \mid \mathscr{F}_t\Big).
\end{aligned}
$$

By using Fatou's lemma and assumption 1. Of Theorem 2 we get

$$
\begin{aligned}
Y_t + \mathbb{E}\Big(Y^-_{\lambda\wedge\sigma^*_t} \mid \mathscr{F}_t\Big) &\leq Y_t + \liminf_n \mathbb{E}\Big(Y^-_{\lambda\wedge\sigma^*_t\wedge\upsilon^+_n} \mid \mathscr{F}_t\Big)\\
&\leq \liminf_n \mathbb{E}\Big(Y^+_{\lambda\wedge\sigma^*_t\wedge\upsilon^+_n} \mid \mathscr{F}_t\Big)\\
&\leq \mathbb{E}\Big(Y^+_{\lambda\wedge\sigma^*_t} \mid \mathscr{F}_t) + \liminf_{n\to+\infty} a^+_n \mathbb{E}\Big(1_{\lambda\wedge\sigma^*_t>\upsilon^+_n} \mid \mathscr{F}_t\Big)\\
&\leq \mathbb{E}\Big(Y^+_{\lambda\wedge\sigma^*_t} \mid \mathscr{F}_t) + \liminf_{n\to+\infty} a^+_n \mathbb{E}\Big(1_{\{\sup_{s\leq T} F(U_s)^+>a^+_n\}} \mid \mathscr{F}_t\Big)\\
&\leq \mathbb{E}\Big(Y^+_{\lambda\wedge\sigma^*_t} \mid \mathscr{F}_t\Big),
\end{aligned}
$$

where we have used the limit appeared in (4).

It follows that for every stopping time $\lambda \in \mathcal{M}_{t,T}$

$$\begin{aligned} Y_t &\leq \mathbb{E}\Big(Y^+_{\lambda\wedge\sigma^*_t} \mid \mathcal{F}_t\Big) - \mathbb{E}\Big(Y^-_{\lambda\wedge\sigma^*_t} \mid \mathcal{F}_t\Big) \\ &= \mathbb{E}\Big(Y_{\lambda\wedge\sigma^*_t} \mid \mathcal{F}_t\Big). \\ &\leq \mathbb{E}\Big(F(U_\lambda)1_{\{\lambda<\sigma^*_t\}} + F(L_{\sigma^*})1_{\{\lambda>\sigma^*_t\}} + F(Q_{\sigma^*})1_{\{\sigma^*_t=\lambda<T\}} + F(\xi)1_{\{\sigma^*_t=\lambda=T\}} \mid \mathcal{F}_t\Big) \\ &= \mathbb{E}\Big(J(\lambda, \sigma^*_t) \mid \mathcal{F}_t\Big). \end{aligned} \tag{6}$$

In force of inequalities (5) and (6) we obtain that for all $\sigma, \lambda \in \mathcal{M}_{t,T}$

$$\mathbb{E}\Big(J(\lambda^*_t, \sigma) \mid \mathcal{F}_t\Big) \leq Y_t = \mathbb{E}\Big[J(\lambda^*_t, \sigma^*_t) \mid \mathcal{F}_t\Big] \leq \mathbb{E}\Big(J(\lambda, \sigma^*_t) \mid \mathcal{F}_t\Big).$$

Then it is immediately checked that

$$\begin{aligned} \inf_{\lambda\in\mathcal{M}_{t,T}} \sup_{\sigma\in\mathcal{T}_t} \mathbb{E}\Big(J(\lambda, \sigma) \mid \mathcal{F}_t\Big) &\leq \sup_{\sigma\in\mathcal{T}_t} \mathbb{E}\Big(J(\lambda^*_t, \sigma) \mid \mathcal{F}_t\Big) \\ &\leq Y_t = \mathbb{E}\Big[J(\lambda^*_t, \sigma^*_t) \mid \mathcal{F}_t\Big] \\ &\leq \inf_{\lambda\in\mathcal{M}_{t,T}} \mathbb{E}\Big(J(\lambda, \sigma^*_t) \mid \mathcal{F}_t\Big) \\ &\leq \sup_{\sigma\in\mathcal{M}_{t,T}} \inf_{\lambda\in\mathcal{M}_{t,T}} \mathbb{E}\Big(J(\lambda, \sigma) \mid \mathcal{F}_t\Big). \end{aligned}$$

Since $\sup_{\sigma\in\mathcal{M}_{t,T}} \inf_{\lambda\in\mathcal{M}_{t,T}} \mathbb{E}\Big(J(\lambda, \sigma) \mid \mathcal{F}_t\Big) \leq \inf_{\lambda\in\mathcal{M}_{t,T}} \sup_{\sigma\in\mathcal{M}_{t,T}} \mathbb{E}\Big(J(\lambda, \sigma) \mid \mathcal{F}_t\Big)$, we have

$$\begin{aligned} Y_t &= \mathbb{E}\Big[J(\lambda^*_t, \sigma^*_t) \mid \mathcal{F}_t\Big] \\ &= \sup_{\sigma\in\mathcal{M}_{t,T}} \mathbb{E}\Big[J(\lambda^*_t, \sigma) \mid \mathcal{F}_t\Big] = \inf_{\lambda\in\mathcal{M}_{t,T}} \mathbb{E}\Big[J(\lambda, \sigma^*_t) \mid \mathcal{F}_t\Big] \\ &= \inf_{\lambda\in\mathcal{M}_{t,T}} \sup_{\sigma\in\mathcal{M}_{t,T}} \mathbb{E}\Big[J(\lambda, \sigma) \mid \mathcal{F}_t\Big] = \sup_{\sigma\in\mathcal{M}_{t,T}} \inf_{\lambda\in\mathcal{T}_t} \mathbb{E}\Big[J(\lambda, \sigma) \mid \mathcal{F}_t\Big], \end{aligned}$$

Theorem 2 is then proved. □

Remark 1 We should remark here that:

1. If $F(L)$ and $F(U)$ are L^1-integrable, i.e. $\mathbb{E}\sup_{t\leq T}(|F(U_t)| + |F(L_t)|) < +\infty$, then the assumption of Theorem 2 are satisfied.

2. If we suppose that $F(x) = e^{\theta x}$ (or $F(x) = -e^{-\theta x}$), $\theta > 0$, we have an utility function which is of exponential type and then our result can give, in particular, a solution to the existence a saddle point for the risk-sensitive problem (see [7] for more details).

4 American Game Option

4.1 Problem Formulation

We deal with American game option or a game contingent claim which is a contract between a seller A and a buyer B at time $t = 0$ such that both have the right to exercise at any stopping time before the maturity time T . If the buyer exercises at time t he receives the amount $L_t \geq 0$ from the seller and if the seller exercises at time t before the buyer he must pay to the buyer the amount $U_t \geq L_t$ so that $U_t - L_t$ is viewed as a penalty imposed on the seller for cancellation of the contract. If both exercise at the same time t before the maturity time T then the buyer may claim Q_t and if neither have exercised until the expiry time T then the buyer may claim the amount ξ. In short, if the the seller decides to exercise at a stopping time $\lambda \leq T$ and the buyer exercises at a stopping time $\sigma \leq T$ then the former pays to the latter the amount:

$$J^1(\lambda, \sigma) = U_\lambda 1_{\{\lambda<\sigma\}} + L_\sigma 1_{\{\lambda>\sigma\}} + Q_\sigma 1_{\{\sigma=\lambda<T\}} + \xi 1_{\{\sigma=\lambda=T\}}.$$

Such game option is considered in a standard securities market consisting of a non-random component S_t^0 representing the value of a savings account at time t with an interest rate r and of a random component S_t representing the stock price at time t. More precisely and following the same idea as in Hamadène [9], we consider a security market $\mathscr{M}$ that contains, say, one bond and one stock and we suppose that their prices are subject to the following system of stochastic differential equations:

$$\begin{cases} dS_t^0 = rS_t^0 dt, \quad S_0^0 > 0 \\ dS_t = S_t(bdt + \delta dB_t), \quad S_0 > 0. \end{cases}$$

Let X be an $\mathscr{F}_t$-measurable random variable such that $X \geq 0$. The classical approach suggests that valuation of options should be based on the notions of a self-financing portfolio and on hedging. For this reason, we give the following definitions.

Definition 3 A self-financing portfolio after t with endowment at time t is X, is a $\mathscr{P}$-measurable process $\pi = (\beta_s, \gamma_s)_{t\leq s\leq T}$ with values in $\mathbb{R}^2$ such that:
(i) $\int_t^T (\mid \beta_s \mid + (\gamma_s S_s)^2) ds < \infty$.
(ii) If $\Delta_s^{\pi,X} = \beta_s S_s^0 + \gamma_s S_s, \quad s \leq T,$ then $\Delta_s^{\pi,X} = X + \int_t^s \beta_u dS_u^0 + \int_t^s \gamma_u dS_u$, $\forall s \leq T$.

Definition 4 A hedge against the game with payoff

$$J^1(s,\lambda) := U_\lambda 1_{\{\lambda<s\}} + L_s 1_{\{s<\lambda\}} + Q_s 1_{\{s=\lambda<T\}} + \xi 1_{\{s=\lambda=T\}},$$

after t whose endowment at t is X is a pair (π, λ), where π is self-financing portfolio after t whose endowment at t is X and a stopping time $\lambda \in \mathscr{M}_{t,T}$, satisfying: P-a.s. $\forall s \in [t, T]$,

$$\Delta^{\pi,X}_{s\wedge\lambda} \geq J^1(s,\lambda).$$

Definition 5 The fair price of a contingent claim game is the infimum of capitals X for which the hedging strategy exists. It is defined by

$$V_t := \inf\{X \geq 0,\ \exists(\pi,\lambda) \text{ such that } \Delta^{\pi,X}_{s\wedge\lambda} \geq J^1(s,\lambda),\ \forall t \leq s \leq T,\ P-a.s.\}.$$

4.2 Fair Price of the Game as a Solution of BSDE with Two Reflecting Barriers

Now, let P^* be the probability on $(\Omega, \mathscr{F})$ under which the actualized price of the asset is a martingale, i.e.

$$\frac{dP^*}{dP} := exp\left(-\delta^{-1}(b-r)B_t - \frac{1}{2}(\delta^{-1}(b-r))^2 t \right), \quad t \leq T.$$

Hence the process $W_t = B_t + \delta^{-1}(b-r)t$ is an $(\mathscr{F}_t, P^*)$-Brownian motion.

Let ξ, L, U and Q be as in the beginning such that: $0 \leq L \leq U$. Assume moreover that assumption (**A.1**) holds true and consider, on the probability space $(\Omega, \mathscr{F}, P^*)$, the following BSDE with two reflecting barriers whose solution exists according to Theorem 1

$$\begin{cases} (i)\ \ Y_t = e^{-rT}\xi + \int_t^T dK_s^+ - \int_t^T dK_s^- - \int_t^T Z_s dW_s,\ t \leq T, \\ (ii)\ \forall t \leq T,\ e^{-rt}L_t \leq Y_t \leq e^{-rt}U_t, \\ \qquad \int_0^T (Y_t - e^{-rt}L_t)dK_t^+ = \int_0^T (e^{-rt}U_t - Y_t)dK_t^- = 0, \text{ a.s.}, \\ (iv)\ Y \in \mathscr{C} \quad K^+, K^- \in \mathscr{K} \quad Z \in \mathscr{L}^{2,d}, \\ (v)\ \ dK^+ \perp dK^-. \end{cases} \tag{7}$$

Let ϱ_t^* and ϑ_t^* be the stopping times defined as follows:

$$\varrho_t^* = \inf\{s \geq t\ :\ Y_s = e^{-rt}U_s\} \wedge T \quad \text{and} \quad \vartheta_t^* = \inf\{s \geq t\ :\ Y_s = e^{-rt}L_s\} \wedge T.$$

If we suppose that $\liminf_{r\to+\infty} rP^*(\sup_{s\leq T} U_s > r) = 0$, it follows then from Theorem 2, since $L \geq 0$, that for all $\sigma, \lambda \in \mathscr{M}_{t,T}$, Y_t solution of BSDE (7) is given by

$$\begin{aligned} Y_t &= \mathbb{E}^*\Big[\overline{J}(\varrho_t^*, \vartheta_t^*) \mid \mathscr{F}_t\Big] \\ &= \inf_{\lambda\in\mathscr{M}_{t,T}} \sup_{\sigma\in\mathscr{M}_{t,T}} \mathbb{E}^*\Big[\overline{J}(\lambda,\sigma) \mid \mathscr{F}_t\Big] = \sup_{\sigma\in\mathscr{M}_{t,T}} \inf_{\lambda\in\mathscr{T}_t} \mathbb{E}^*\Big[\overline{J}(\lambda,\sigma) \mid \mathscr{F}_t\Big], \end{aligned} \tag{8}$$

where

$$\overline{J}(\lambda,\sigma) = e^{-r\lambda}U_\lambda 1_{\{\lambda<\sigma\}} + e^{-r\sigma}L_\sigma 1_{\{\lambda>\sigma\}} + e^{-r\lambda}Q_\sigma 1_{\{\sigma=\lambda<T\}} + e^{-rT}\xi 1_{\{\sigma=\lambda=T\}}.$$

The main result of this section is the following.

Theorem 3 *Assume that* $\liminf_{r\to+\infty} rP^*(\sup_{s\le T} U_s > r) = 0$. *Then, the fair price of our game is given by* $V_t = e^{rt}Y_t$, *for any* $t \le T$. *Moreover, a hedge after* t *against the option exists and it is given by:*

$$\gamma_s = \frac{e^{rs}Z_s}{\delta S_s} 1_{\{s\le\vartheta_t^*\}} \text{ and } \beta_s = \left(e^{rs}(Y_t + \int_t^s Z_u dW_u) - \gamma_s S_s\right)(S_s^0)^{-1}, \ \forall s \in [t,T].$$

Proof Let (π,λ) a hedge after t against the option. Therefore $\lambda \in \mathscr{M}_{t,T}$ and $\pi = (\beta_s,\gamma_s)_{t\le t\le T}$ is a self-financing portfolio whose value at t is X satisfying $\Delta^{\pi,X}_{s\wedge\lambda} \ge J^1(s,\lambda), \ \forall t \le s \le T$. But

$$e^{-r(s\wedge\lambda)}\Delta^{\pi,X}_{s\wedge\lambda} = e^{-rt}X + \delta\int_t^{s\wedge\lambda} \gamma_u S_u e^{-ru} dW_u \ge e^{-r(s\wedge\lambda)}J^1(s,\lambda), \ \forall t \le s \le T.$$

Let $\sigma \ge t$ be a stopping time. Putting $s = \sigma$ and taking the conditional expectation we obtain

$$e^{-rt}X \ge \mathbb{E}^*\Big(e^{-r(\sigma\wedge\lambda)}J^1(\sigma,\lambda) \mid \mathscr{F}_t\Big).$$

In view of relation (8) we have

$$\begin{aligned} e^{-rt}X &\ge \sup_{\sigma\in\mathscr{M}_{t,T}} \mathbb{E}^*\Big(e^{-r(\sigma\wedge\lambda)}J^1(\sigma,\lambda) \mid \mathscr{F}_t\Big) \\ &\ge \inf_{\lambda\in\mathscr{M}_{t,T}} \sup_{\sigma\in\mathscr{M}_{t,T}} \mathbb{E}^*\Big(e^{-r(\sigma\wedge\lambda)}J^1(\sigma,\lambda) \mid \mathscr{F}_t\Big) \\ &= \inf_{\lambda\in\mathscr{M}_{t,T}} \sup_{\sigma\in\mathscr{M}_{t,T}} \mathbb{E}^*\Big(\overline{J}(\sigma,\lambda) \mid \mathscr{F}_t\Big) \\ &= Y_t. \end{aligned}$$

Henceforth $V_t \ge e^{rt}Y_t$. Let us now prove the converse inequality. It follows that for every $t \le s \le T$,

$$
\begin{aligned}
& Y_t + \int_t^{s\wedge\vartheta_t^*} Z_u dW_u \\
& \le Y_{s\wedge\vartheta_t^*} - \int_t^{s\wedge\vartheta_t^*} dK_u^- \\
& \le Y_{s\wedge\vartheta_t^*} \\
& \le e^{-rs} U_s 1_{\{s<\vartheta_t^*\}} + e^{-r\vartheta_t^*} L_{\vartheta_t^*} 1_{\{s>\vartheta_t^*\}} + e^{-rs} Q_{\vartheta_t^*} 1_{\{\vartheta_t^*=s<T\}} + e^{-rT}\xi 1_{\{\vartheta_t^*=s=T\}} \\
& = e^{-r(s\wedge\vartheta_t^*)} J^1(s, \vartheta_t^*).
\end{aligned}
$$

Hence for every $t \le s \le T$,

$$
J^1(s, \vartheta_t^*) \ge e^{r(s\wedge\vartheta_t^*)}(Y_t + \int_t^{s\wedge\vartheta_t^*} Z_u dW_u).
$$

Now if we put for all $s \in [t, T]$, $\gamma_s = \frac{e^{rs} Z_s}{\delta S_s} 1_{\{s\le\vartheta_t^*\}}$ and $\beta_s = \Big(e^{rs}(Y_t + \int_t^s Z_u dW_u) - \gamma_s S_s\Big)(S_s^0)^{-1}$.

Hence $(\beta_s, \gamma_s)_{t\le s\le T}$ is a self-financing portfolio whose value at t is $e^{rt}Y_t$. On other hand we have

$$
e^{r(s\wedge\varrho_t^*)}(Y_t + \int_t^{s\wedge\varrho_t^*} Z_u dW_u) \ge J^1(s, \varrho_t^*), \ \forall s \in [t, T].
$$

Hence $((\beta_s, \gamma_s)_{t\le t\le T}, \varrho_t^*)$ is a hedge against the game option. Then $e^{rt}Y_t \ge V_t$. Henceforth $e^{rt}Y_t = V_t$. The proof of Theorem 3 is then achieved. □

References

1. Alario-Nazaret, M., Lepeltier, J.P., Marchal, B.: Dynkin Games. Lecture Notes in Control and Information Sciences, vol. 43, pp. 23–42. Springer, Berlin (1982)
2. Bensoussan, A., Friedman, A.: Non-linear variational inequalities and differential games with stopping times. J. Funct. Anal. **16**, 305–352 (1974)
3. Bismut, J.M.: Sur un problème de Dynkin. Z. Wahrsch. Verw. Gebiete **39**, 31–53 (1977)
4. Cvitanic, J., Karatzas, I.: Backward SDEs with reflection and Dynkin games. Ann. Prob. **24**(4), 2024–2056 (1996)
5. Dumitrescu, R., Quenez, M.C., Sulem, A.: Generalized Dynkin games and doubly reflected BSDEs with jumps. arXiv:1310.2764v2 (2014)
6. Dynkin, E.B., Yushkevich, A.A.: Theorems and Problems in Markov Processes. Plenum Press, New York (1968)
7. El-Karoui, N., Hamadène, S.: BSDEs and risk-sensitive control, zero-sumand nonzero-sum game problems of stochastic functional differential equations. Stoch. Process. Appl. **107**, 145–169 (2003)
8. Essaky, E.H., Hassani, M.: Generalized BSDE with 2-reflecting barriers and stochastic quadratic growth. J. Differ. Equ. **254**(3), 1500–1528 (2013)
9. Hamadène., S.: Mixed zero-sum stochastic differental game and American game options. SIAM J. Optim. **45**(2), 496–518 (2007)
10. Hamadène, S., Lepeltier, J.P.: Reflected BSDEs and mixed game problem. Stoch. Process. Appl. **85**, 177–188 (2000)

11. Hamadène, S., Hassani, M.: BSDEs with two reflecting barriers: the general result. Prob. Theory Relat. Fields **132**, 237–264 (2005)
12. Hamadène, S., Hassani, M.: BSDEs with two reflecting barriers driven by a Brownian and a Poisson noise and related Dynkin game. Electron. J. Prob. **11**(5), 121–145 (2006)
13. Hamadéne, S., Zhang, J.: The continuous time nonzero-sum Dynkin game and application in game options. SIAM J. Control Optim. **48**(5), 3659–3669 (2010)
14. Karatzas, I., Li, Q.: BSDE approach to non-zero-sum stochastic differential games of control and stopping. Submitted (2011)
15. Kifer, Y.: Game options. Finance Stoch. **4**, 443–463 (2000)
16. Lepeltier, J.P., Maingueneau, M.A.: Le jeu de Dynkin en théorie générale sans l'hypothèse de Mokobodski. Stochastics **13**, 25–44 (1984)
17. Morimoto, H.: Dynkin games and martingale methods. Stochastics **13**, 213–228 (1984)
18. Neveu, J.: Discrete-Parameter Martingales. North-Holland, Amsterdam (1975)
19. Touzi, N., Vieille, N.: Continuous-time-Dynkin games with mixed strategies. SIAM J. Control Optim. **41**(4), 1073–1088 (2002)

Erratum to: Statistical Methods and Applications in Insurance and Finance

M'hamed Eddahbi, El Hassan Essaky and Josep Vives

Erratum to:
M. Eddahbi et al. (eds.), *Statistical Methods and Applications in Insurance and Finance*, Springer Proceedings in Mathematics & Statistics 158, DOI 10.1007/978-3-319-30417-5

The book was inadvertently published with an incorrect subtitle as CIMPA School, Marrakech and **El Kelaa M'gouna**, Morocco, April 2013. The correct subtitle is CIMPA School, Marrakech and **Kelaat M'gouna**, Morocco, April 2013.

The updated original online version for this book frontmatter can be found at
DOI 10.1007/978-3-319-30417-5

M. Eddahbi (✉)
Department of Mathematics, Université Cadi Ayyad, Marrakesh, Morocco
e-mail: m.eddahbi@uca.ma

E.H. Essaky
Faculté Polydisciplinaire de Safi, Université Cadi Ayyad, Safi, Morocco
e-mail: essaky@uca.ma

J. Vives
Faculty of Mathematics, University of Barcelona, Barcelona, Spain
e-mail: josep.vives@ub.edu

M. Eddahbi et al. (eds.), *Statistical Methods and Applications in Insurance and Finance*, Springer Proceedings in Mathematics & Statistics 158,
DOI 10.1007/978-3-319-30417-5_10

The manufacturer's authorised representative in the EU is Springer Nature Customer Service Centre GmbH, Europaplatz 3, 69115 Heidelberg, Germany. If you have any concerns regarding our products, please contact ProductSafety@springernature.com

Printed and bound by CPI Group (UK) Ltd, Croydon, CR0 4YY
30/09/2026
02225643-0006